Opening Prayer

All praise is due to Allah, the Beneficent, the Gracious
He Who created the universe and made the Earth spacious
He is the Most Noble and taught with the pen
And put knowledge into the hearts of men
And prayer and peace upon His Messenger, the Perfect
And upon his family and Companions, the Elect
And upon his followers until the end of time
Until they see the Face of the One Who is Sublime.

Title: CONCERNING DIVINE WISDOM IN THE CREATION OF MAN

ISBN: 978-1-952306-48-8 (PAPERBACK)

978-1-952306-44-0 (HARDCOVER)

FIRST EDITION | NOVEMBER 2022

Author: ABŪ ḤĀMID AL-GHAZĀLĪ

Translation & Commentary:
SHEIKH MOHAMMED AMIN KHOLWADIA
DR. KAMRAN M. RIAZ

Proofreading: WORDSMITHS

Typesetting: IGPCONSULTANTS

www.imamghazali.co

Concerning Divine Wisdom in the Creation of Man

The Proof of Islam

Abū Ḥāmid al-Ghazālī ﷺ
(1058-1111)

Translation by
Dr. Kamran M. Riaz

Commentary by
Sheikh Mohammed Amin Kholwadia
Dr. Kamran M. Riaz

Jointly Published by

Imam Ghazali Publishing
Darul Qasim
In the quest for more knowledge

Contents

Acknowledgements

UNDOUBTEDLY, THIS TRANSLATION PROJECT was a monumental and challenging task that I still feel unworthy and unqualified to have embarked upon. When Sheikh Amin urged me to commence this work nearly ten years ago, I agreed to do so contingent on his consent to mentor, correct, and supplicate for me every step of the way. I am therefore most grateful to Sheikh Amin for providing the idea for this project, his assistance with the preparation of this translation with such diligence that rivals the diligence of a mother in caring for her infant, and his illuminating commentary that will hopefully reintroduce a work of Ḥujjat al-Islām to an English-speaking audience; particularly among those in the medical and healthcare fields.

From the Darul Qasim faculty, I am very thankful to Dr. Choukri Heddouchi, Ph.D., for his critical review of the translation, especially in improving the sophistication and flow of the English text while maintaining the essence of the Arabic text. A special recognition to Dr. Ahsan Arozullah for his excellent foreword to the text, providing a reference point for medical professionals, and eloquently capturing the feelings of many Muslim medical students as they progress through their studies. I am also appreciative of Mawlānā Samir Ali for his meticulous research and academic cross-referencing of the various Qur'anic verses and Prophetic hadiths. I would also like to thank the Darul Qasim Bioethics Group – particularly Dr. Umar Shakur DO and Dr. Yasir Akhtar MD for their perspective, comments, and suggestions to improve the quality of the text. Darul Qasim is truly a special place filled with extraordinary people. I pray that it continues to be a source of benefit to future generations, remaining in the land long after we have departed, "*...as for that which benefits mankind, it remains in the land.*"[1]

To everyone at Imam Ghazali Publishing, especially Muhammad Adnaan Sattaur: this book would not have seen light of day had it not been for your team's help. May Allah reward you, your family, and everyone at the Publishing House with the best of both worlds. There are many other people who have made *duʿā* for this project and assisted in various ways over the years – big and small, remembered and [sadly] unremembered – that I also want to recognize. Many of you prefer anonymity, so I will instead attempt to continuously keep you and your loved ones in my *duʿās*.

Finally, I am grateful to my family for their support and patience over these years; especially my wife, for providing me with the time and freedom to finish this project. To my brother, Dr. Hafiz Rehan Riaz MD: I am blessed to have you as my younger brother and my biggest critic. Your patience and critical review of this project over the past ten years is something that I will cherish forever. Most importantly, I dedicate this work to my mother, who instilled in me a desire to study every form of knowledge and introduced me to *Ḥujjat al-Islām* when I was a child. I pray that whatever reward I get from this book is conveyed to you instead, many times over.

Dr. Kamran Riaz
Darul Qasim

---

1 *al-Raʿd*, 17.

Foreword

AHSAN AROZULLAH, M.D., M.P.H.

THE MASTERY OF HUMAN ANATOMY has come to represent a distinct rite of passage in the hallowed halls of the medical profession. Anatomy is typically a required course in the first year of medical school. For centuries, young medical students have been required to study the anatomy of the human body as a foundation of their medical education. Physicians may recall the nervous anticipation of initially encountering a cadaver in the dissection room, and the stress of preparing for the final practical anatomy examination. Medical students quickly learn that the one who passes the anatomy course has a high likelihood of ultimately becoming a physician.

The study of anatomy for the Muslim medical student serves as a potential means to fulfill two distinct goals: to become a physician with knowledge of diseases and cures, and to increase one's attachment to the Creator through a deeper appreciation of the details of His creation. The challenging question that naturally arises is whether the Muslim medical student can successfully achieve these two goals during their medical education. The assumption that multi-tasking – performing more than one *action* at a time – can improve efficiency is predicated on the notion that the multiple actions undertaken are aligned towards accomplishing *one focused goal.* Conversely, man is limited in his ability to focus on accomplishing more than one *goal* at a time, even when working on only one *action* at a time. This human limitation, although rarely discussed in the context of medical education, should not be surprising to the Muslim physician.

Muslim physicians may recall that, during their time in medical school, they hoped that the study of anatomy would mark the beginning of a journey towards a deeper appreciation of Allah's creativity. By extension, they sought to enhance their own personal relationship with Allah. They may also remember that their focus on admiring Allah's creativity was often diverted by the exhaustive and detailed preparation required to pass anatomy examinations. By the end of the anatomy course, the initial hope of deepening one's relationship with Allah through studying anatomy was often replaced with fleeting moments of awareness of Allah's creativity.

The scientific paradigm currently defines the lens through which medical students learn human anatomy. This paradigm presents several challenges to the Muslim medical student seeking to acquire *religious* knowledge through the study of anatomy. The scientific paradigm asserts that the only valid sources of knowledge are experiences occurring through the physical senses (empiricism), and that man's life is therefore likely limited to the body's physical life in this world. In contrast, the Islamic paradigm defines revelation – the Qur'an and Sunnah – as the primary source of knowledge. Based on revelation, the Islamic paradigm asserts that man's life is not limited to the body's physical life in this world. The human anatomy textbooks currently used in medical education are almost exclusively based on the scientific paradigm. Therefore, a Muslim medical student is faced with a daunting challenge: where among the scientific paradigm's underlying assumptions is there an opening for an *Islamic* study of anatomy?

The purpose of Islam is to facilitate one's salvation in the afterlife, through which the greatest reward will be to meet Allah, one's Creator and Cherisher. Islam also provides a paradigm that the Muslim medical student may apply when studying human anatomy. The Islamic paradigm expands man's existence beyond the physical body of this world by considering existence prior to birth into this world and existence continuing into the afterlife. The understanding of the role of the physical body in each stage of man's existence provides an Islamic lens through which human anatomy can be studied, together with its implications. In the study of anatomy, the Islamic paradigm does not preclude the use of the physical senses and the mind as sources of knowledge. The purpose of Divine Revelation is to not preclude the physical senses – including the mind – as sources of knowledge. Knowledge of Revelation, through the Qur'an and the Sunnah, adds value to these two sources of knowledge.

The mission of Darul Qasim is to facilitate the delivery and transmission of knowledge derived from the Prophethood of Muhammad ﷺ by utilizing the Islamic paradigm. Through the translation and commentary of this treatise on the human anatomy by Imam al-Ghazālī – a master of the Islamic paradigm – Darul Qasim provides the seeker of religious knowledge a window into Allah's creativity through the *Islamic* study of anatomy. The broad audience of medical students, physicians, healthcare professionals, as well as non-medical seekers of religious knowledge will likely appreciate the details of human anatomy described by Imam al-Ghazālī through an Islamic lens. They will also recognize that these details reflect a distinctly different paradigm from the scientific lens through which they have previously studied anatomy.

The goal of Darul Qasim in translating this work into English is to facilitate and rekindle a desire among Muslim healthcare professionals to seek an Islamic study of medical sciences. Although the Muslim medical professional initially studies human anatomy through a scientific lens, this does not preclude the opportunity to undertake a religious study of medical sciences. In addition, these studies should be undertaken with the sole purpose of appreciating Allah's creativity and work towards synthetizing both the Islamic and scientific paradigms. However, just as medical students depend on their professors (who are grounded in the scientific paradigm) to facilitate their medical education, students will likely benefit from formal study with teachers grounded in the Islamic paradigm when studying the wonders of human anatomy.

Introduction to Muslim Epistemology: Prerequisites

SHEIKH MOHAMMED AMIN KHOLWADIA

ALL PRAISE IS DUE TO ALLAH, the Eternally Beneficent, Eternally Merciful.

O Allah! Send blessings upon Muhammad ﷺ: Your servant, Your Messenger, and Your unlettered Prophet. Send blessings upon his trustworthy Companions, his family members, his offspring, and his followers until the Last Day.

THE STATUS OF AL-GHAZĀLĪ IN MUSLIM ACADEMIA

Imam Abū Ḥāmid Muhammad ibn Muhammad al-Ghazālī of Ṭūs (a city in modern-day Iran) is universally acclaimed as one of the most outstanding scholars of Sunni Islam. His writings, such as his masterpiece on the principles of Islamic jurisprudence, *al-Mustaṣfā*, are still widely read and discussed. He also wrote on Islamic dogma and theology, including the famous *al-Munqidh min al-Ḍalāl* and the controversial *Tahāfut al-Falāsifah*. His magnum opus (*Iḥyā' ʿUlūm al-Dīn*) and his *Kīmiyā-yi Sa'ādat* are without doubt two of the most brilliant works on Islamic ethics and metaphysics.

Through his writings, al-Ghazālī displayed great mastery in Islamic disciplines in both expression and content; in image and reality; in words and meaning; and in body and spirit. However, despite the unique prestige that al-Ghazālī still enjoys in academia, there are scholars who find him difficult to read, and claim that his writings are full of apparent contradictions. They claim that al-Ghazālī does not seem to offer any consistent foundation for some of his doctrinal positions. Some argue that this is because we cannot accurately date his writings, thus preventing us from appropriating any development or change in his thought process. Others contend that this is due to the fact that al-Ghazālī borrowed heavily from the language of Muslim philosophers, and was therefore unable to develop a language that best suited his views. Some observers went as far as denouncing him for engaging in polemics in the first place.

In general, these critics have many questions:

Was al-Ghazālī in favor of learning hard sciences?
Was he against everything philosophers stood for?
Was he a diehard Ash ʿarī[2]*?*
Did he lean towards a more neutral position in theology?
Was his mystical writing a mirror of his legal philosophy? Or vice versa?

These are some of the issues raised by critics after reading the works of this great Muslim thinker and theologian. We should add to this list of questions a more foundational question that can be based on the following premise.

Like all other Muslim scholars, al-Ghazālī adhered to an Islamic

---

2 A Sunni theological school of thought named after Imam al-Ashʿarī, who broke off from his Muʿtazilite teacher and became a vehement opponent of Muʿtazilite polemics. This is briefly discussed in the next few pages.

academic foundation in his theistic and theological writings. He believed in the existence of the One Divine Being (Allah), the Prophethood of Muhammad ﷺ, and the resurrection on the Last Day. Accepting these maxims has always been regarded as a prerequisite to enter the discourse of Islamic sciences. Conversely, only someone who has explicitly denied any one of these three doctrines as essential Islamic dogma was seen as a non-Muslim. It is precisely because of this vast common ground that orthodox Muslim scholars did not go out of their way to brand unorthodox scholars as outright non-believers or apostates (*murtadd*). Islamic law vis-à-vis apostasy (in the Sunni tradition) applies only after someone has been convicted of outrageous and obvious blasphemy. Islamic law seeks to include – but not encourage – theological aberrations as much as possible.

The question at hand is whether Muslim theologians gave such latitude to religious (dogmatic) pluralism. If al-Ghazālī[3] also believed that Muslims should not adopt any mode of ecclesiastical sentencing within the Muslim community, then why was he so adamant against Muslim philosophers, such that he felt compelled to write an entire book against them?[4] Why did he feel compelled to vehemently argue against scholars who were not quite so orthodox as him? Did he feel that these philosophers – who were mostly from the Muʿtazilite sect – crossed the boundaries of acceptable theological views and consequently fell into kufr (disbelief)?

We hope to address these quandaries by evaluating the epistemology of al-Ghazālī. We believe that unless Muslim scholars

3 Al-Ghazālī wrote on this subject in a work entitled: *Fayṣal al-Tafriqah Bayn al-Islām wa al-Zandaqah* (*The Distinctive Criterion to Distinguish Islam from Masked Infidelity*).

4 This is the well-known *Tahāfut al-Falāsifah* (*The Incoherence of the Philosophers*).

present a holistic exposé of Sunni epistemology, we will not be able to appreciate the work of Sunni scholars. If we do present such an exposé, we will be able to see why al-Ghazālī used the language of Muslim philosophers without becoming a party to their theories, as suggested by certain scholars.[5] On the whole, Sunni scholars were never averse to coining new terms in order to express old ideas and theories. For example, the science of *'ilm al-rijāl* (the standards by which Sunni scholars measure the integrity and reliability of Ḥadīth transmitters) was not at all in vogue during the time of the Companions (Ṣaḥābah) of the Prophet Muhammad ﷺ, but became a normative science by the end of the 1st century AH.

In short, we believe that al-Ghazālī should not be condemned for using a language that was different in form than the one used by traditional Muslim scholars. If he was guilty of that, then so were the traditional scholars who studied and taught Ḥadīth. Scholars of Ḥadīth classify several narrations from the Prophet ﷺ as being "weak" (*ḍaʿīf*). If an unlearned simpleton suggested that Ḥadīth scholars did not respect their Prophet since they ascribe weakness to his statements, we would all condemn him for his ignorance! We would instruct him to move outside of his operational knowledge of Islam and seek academic familiarity with the Islamic sciences.

Likewise, it is our understanding that reading the works of al-Ghazālī – and those of every other scholar – requires learning and understanding the language and idioms with which the scholars used. Upon learning this technical language – which is required for reading any academic discipline (such as medicine) – the student

5 This was the approach of classical Muslim polemists (*mutakallimūn*), who used Hellenic language and methods to disprove heretical theories held by Muslim philosophers. Ibn Taymiyyah and his supporters disapproved of the use of such language and called to rid the Muslim community of such polemical discussions.

will appreciate the paradigms and theories upon which al-Ghazālī bases his arguments.

THE SUNNI THEORY OF ISLAM

As an institute for higher Sunni learning, Darul Qasim aims to express a theory of Islam – as held by early Sunni scholars – in its quest to represent a holistic view of Sunni scholarship. One of our goals is to publish works on essential paradigmatic principles based on the teachings of early Muslim theorists. The present work represents the first of our efforts, and Allah is the One Who reconciles our efforts with His Acceptance.

It is our reading at Darul Qasim that orthodox Sunni scholars who came after the Companions ﷺ did not initiate any new primary doctrinal and theological maxims (*ʿaqāʾid*). They either enhanced or refined maxims that were previously adopted or debated. In their efforts, they set out to use a different type of language and terminology. This was designed not to distort the content of Sunni Islam, but rather to bolster its academic and theoretical consistency in a society that was increasingly being stripped of it by undisciplined political parties. They tried – sometimes very violently – to change the pre-established content of Islam in response to the context of political Islam in their times. These zealots, who developed into deviant sects later, used scriptural references to justify their beliefs and agendas.[6] Mainstream Muslim scholars at the time realized the need to give scriptural text a visible coherent structure in order to defend these primary maxims from aberrations and distortions. The need for a new language was quite obvious to Sunni scholars during

6 Specific examples of this will be discussed later on in this introduction.

that era. For example, Imam Muslim mentions this phenomenon in his introduction to his collection of Ḥadīth when he discusses the need for *isnād* (a chain of narrators). It is quite clear today that in order to study the works of Abū Ḥanīfah, Mālik, al-Shāfiʿī, and Aḥmad ibn Ḥanbal ﷺ, one would need to learn their language and terminology. A simplistic familiarity with the Arabic language would be far from adequate for such an enormous task.[7]

In brief, we see that the early Sunni scholars after the Companions ﷺ started to develop an academic language in order to facilitate the learning of Islamic sciences. Their intent was to preserve the religious content of early Islam. Their intent was **not** to introduce any new ideas and theories through their "new" language.

We can see this more clearly when we view Sunni literary history through the lens of Sunni theology. Sunni theology is primarily a faith-based narrative of how the Companions ﷺ of the Prophet Muhammad ﷺ understood and conveyed revelation that came to him.[8] It is our belief that Sunni Islam had crystalized within the lifetime of the major Companions ﷺ of the Prophet ﷺ.

THE ORAL TRADITION AND THE WRITTEN TRADITION

It is evident that the Companions ﷺ played a role in the crystallization of Islam after the departure of the Prophet Muhammad ﷺ. Their role was complementary to the perfection and refinement of Islam itself; they understood their faith and practice based on the convention

7 This is analogous to a high school student who knows the English language and may have taken a course in biology, compared to the medical professional who has mastered the vocabulary and science of anatomy. While there may be common terms that a novice and expert may both appreciate, training in a discipline – especially knowledge of its terminology – is ultimately a key prerequisite for further serious inquiry.

8 As opposed to one that is based on historical and political events.

of an oral tradition. The Prophet ﷺ did not read or write, so Islam was primarily preserved and promoted through the oral tradition. Although many Companions رضي الله عنهم knew how to read and write, the written tradition was not used as a primary convention. They saw the need to document the oral tradition in written form as a *secondary* convention of preserving Islamic dogma. We see this very clearly in the story of how the Qur'an was documented.

The Prophet ﷺ had given the Companions رضي الله عنهم instructions to write down the Qur'anic text. This was achieved by recording the Qur'an on various materials, such as parchment and date palm leaves. This is evident in the popular story of the conversion of ʿUmar ibn al-Khaṭṭāb رضي الله عنه to Islam, that includes an incident wherein he read the Qur'an written on some kind of material.[9] After the departure of the Prophet ﷺ from this world, it was ʿUmar رضي الله عنه who had initially advised Abū Bakr رضي الله عنه – the first *khalīfah* (successor) to the Prophet ﷺ – to take on this task of preserving the Qur'an in a written book. At first, Abū Bakr رضي الله عنه vehemently resisted this idea by arguing that the Prophet ﷺ himself did not commission the writing of the Qur'an into the form of one book. ʿUmar رضي الله عنه was able to persuade Abū Bakr رضي الله عنه by arguing that several *ḥuffāẓ* (memorizers of the Qur'an) had been martyred in the recent Battle of Yamāmah. He argued that preserving the Qur'an in written form was not with the intention of *replacing* the oral tradition, but rather to serve as a further method of preservation.

Abū Bakr رضي الله عنه then commissioned this task to Zayd ibn Thābit رضي الله عنه, the Scribe of the Prophet ﷺ, and one of the best reciters of the Divine Revelation.[10] Zayd رضي الله عنه initially resisted and criticized the two erudite

---

9 Referring to *Surah Ṭāhā*, 20.

10 The entire story of this incident is narrated by Zayd ibn Thābit himself, as mentioned in

Companions for asking him to do something the Prophet ﷺ never did. Abū Bakr ؓ and ʿUmar ؓ persuaded Zayd ؓ by articulating their point of view over a period of time, convincing him to head a commission to complete this laborious task. Zayd ؓ collected any and every written material that the Companions ؓ had written the Qur'an upon. However, he also established a condition: every written verse had to be vetted by at least two Companions who had heard the verse through the oral tradition, that is, from the Prophet ﷺ.[11] For the Companions ؓ, this written collection of Divine Revelation did not undermine the primacy of the oral tradition as a source of acquiring, preserving, and disseminating Islamic knowledge.[12]

Preserving the Qur'an is an explicit Divine Mandate, as evidenced in the well-known verse: "*Indeed, We have revealed the Remembrance, and We will assuredly guard it (from corruption and distortion).*"[13] The Prophet ﷺ executed this mandate by memorizing

---

Ṣaḥīḥ al-Bukhārī (Volume 6, Book 61, hadith #509).

11 Every verse gathered by this commission was vetted by at least two Companions, with the exception of the end of *Surah al-Tawbah*, which was only found with Khuzaymah ibn Thābit al-Anṣārī, who was given the honorific title "The One with Two Witnesses (*Dhū al-shahādatayn*)". Abū Dawūd mentions an incident during the lifetime of the Prophet ﷺ wherein he had bought a horse from a Bedouin. The Bedouin denied the sale and the Prophet ﷺ indicated that he had indeed bought the horse. The Bedouin demanded a witness to corroborate the Prophet's statement. Khuzaymah testified that the Prophet ﷺ had indeed bought the horse. The Prophet ﷺ asked him on what grounds he gave testimony. Khuzaymah replied: "By considering you trustworthy, O Messenger of Allah." The Prophet ﷺ then considered the witness of Khuzaymah equal to the witness of two people.

12 It is noteworthy that modern medicine has a strong history and element of the oral tradition. The first two years in medical school involve lectures and exams (the written tradition), while the final two years require practical training under the guidance and supervision of senior preceptor physicians, who then vet the capability of the medical student (the oral tradition). Board certification exams in many medical disciplines also have both written and oral exam components. The oral tradition is not the dominant source of knowledge in these disciplines, but it certainly plays a major role in conferring certification upon a practitioner.

13 *al-Ḥijr*, 9.

the Qur'an himself and instructing others to do the same. Thus, the Qur'an was preserved first through the oral tradition. Abū Bakr complemented this mandate by preserving the Qur'an further through the written tradition. The Companions introduced the written tradition only after they realized evidence of a precedent in the oral tradition. This was the first collection of the Qur'an in the form of a written book.

Similarly, until the time when ʿUmar became the *khalīfah*, Muslims were reciting the Qur'an in their own respective dialects. In the month of Ramadan, Ubayy ibn Kaʿb, one of the reciters of the Qur'an, led a group of Muslims in a night prayer known as the *tarāwīḥ* prayers. Ubayy led the congregation and offered 20 *rak ʿāt* (units) of prayer by reciting only one single dialect. ʿUmar considered the enactment of 20 *rak ʿāt* of the *tarāwīḥ* prayers during the nights of Ramadan as a good *bid ʿah* (innovation), in the sense that the prayer enabled Muslims to recite in the dialect of a single reciter. The Companions then agreed that reciting the Qur'an in one single method of recitation was complementary to the Divine Intent of the gathering of the Qur'an.[14] They did not see it as a condemned innovation. As a result, the Companions around ʿUmar, including those who were well-recognized as erudite scholars by the Prophet himself, agreed upon this practice unanimously.[15]

But reciting the Qur'an solely in one dialect in prayer is also part of the oral tradition, even though recitation in other dialects was permissible. During the reign of the third *khalīfah* ʿUthmān, the Companions saw the need to officially standardize the

---

14 As per *al-Qiyāmah*, 17: "*Indeed, upon Us is its collection and its recitation.*"

15 This consensus (*ijmāʿ*) amongst the Companions is seen by Sunni scholars as legally and Islamically binding.

recitation of the Qur'an according to the dialect of the Quraysh, which was the dialect of the Prophet ﷺ. Hence, ʿUthmān ؓ – like Abū Bakr ؓ – commissioned Zayd ؓ again to this task based on the precedent that ʿUmar ؓ established, where he allowed Ubayy ؓ to offer *tarāwīḥ* exclusively in one dialect. The written copy of the Qur'an that ʿUthmān ؓ wrote (known as the Muṣḥaf of ʿUthmān) was upheld by the fourth *khalīfah* ʿAlī ؓ, who saw no reason to add to or subtract from the Muṣḥaf. Once again, we see how the Companions ؓ used the written tradition to preserve an oral tradition by playing a complementary role in the crystallization of Islam.

CONTENT VS CONTEXT

Those who followed the Companions ؓ – the Followers (*Tābiʿūn*) – reported this narrative of the oral tradition and exposed the Sunni maxims (*ʿaqā'id*) that the primary **content** of Islamic creed and practice does not change according to or because of **context**. Any need-based change had to be premised on either a primary source of Divine Revelation or an existing precedent. The precedent – in this case, preserving Divine Revelation through the oral tradition – served as a basis for this belief. The antecedent – documenting the Qur'an in a written book – had to be enacted on the back of the precedent and not as fresh legislation or as an innovation (*bidʿah*). The Followers saw how the Companions ؓ required proof in Divine Revelation before they thought about legislating based on public interest and need.

The grand act of preserving the Qur'an in a single dialect in a written document is the reason for the availability of the Qur'an today as a primary source of Islamic knowledge. Early scholars termed this process of using a precedent from which an analogy is drawn for an antecedent as the process of *qiyās* (legal analogy). By giving this legal process a name, the early scholars did not distort Islam; they merely documented what already existed in the oral tradition. *Qiyās* was a term coined by the Followers for a genuine tradition in Islam. We see very clearly that Sunni Muslims were never averse to coining a new language for traditional maxims.

This is an example of how the Companions رضي الله عنهم accommodated a change based on need. What follows is an example where they did not accommodate any change based on context and a perceived need.

PRESERVING THE CONTENT OF ISLAM

Sunni theology does not subscribe to a system of theocracy. The Followers knew that Abū Bakr رضي الله عنه, the first successor of Muhammad ﷺ, did not want Muslims to call him *Khalīfatullāh* (God's deputy), as Muhammad ﷺ is still His Messenger. He instructed them to call him the deputy of the Messenger of Allah (*Khalīfatu Rasūlillāh*). Thus, the Followers saw the government of the Companions رضي الله عنهم as being representative of Muhammad ﷺ – a human being – and not of the Divine. The Followers observed that the Companions رضي الله عنهم guided themselves through the teachings of a perfect human being and did not see this as being contrary to Divine Intent. They also noticed that the Companions رضي الله عنهم followed each other's advice and opinions on Islamic issues that were not discussed during the time of the Prophet ﷺ.[16]

The Companions رضي الله عنهم were the only group of people who could claim that they heard revelation directly from him. Therefore, they did not need to validate their own religious integrity. The Followers concluded that if they desired to procure their salvation, they had to

16 As we have just seen with the compilation of the Qur'an.

follow the lead of the Companions ﷺ in Islamic issues. This was the only method available to them, as revelation had been terminated with the departure of the Prophet ﷺ from this world.

The Followers also knew that the Companions ﷺ had proven, through their impeccable faith in Prophethood, that the meaning of revelation required much more than mere intellectual acumen.[17] They found that the Companions ﷺ relied on revelation in matters of salvation and worship. They also saw that they followed the Prophet ﷺ in the spirit of Islam, which engenders a unique desire to worship Allah. Hence, they saw that the Companions ﷺ were devout in ritual worship (*ʿibādāt*) and sought Divine Assistance in their daily affairs. It was the combination of these three factors – the first-hand witnessing of revelation by being Companions ﷺ, the reliance on revelation in understanding their religion (*fiqh*), and the seeking of Divine Assistance through worship – that convinced the Followers of the merit of the Companions ﷺ above all other Muslims.

These three factors in the Companions ﷺ showed the Followers that the Companions ﷺ possessed an intellectual acumen in Islamic affairs that was unassailable for someone who is not a Prophet. Although Sunnis do not maintain that any individual Companion was infallible, they do believe that the collective understanding of Islam by the Companions ﷺ was guided by the Prophetic light that they received by being with him.[18] Sunnis also believe that this Prophetic light is not an asset that one can acquire after the Messenger ﷺ left this world. Rather, Allah granted this

17 This point is discussed further on in this introduction.

18 Sunni Muslims maintain that this companionship (ṣuḥbah) was a distinctive quality and is what gave the Companions superiority over other Muslims. Hence, they are referred to as Ṣaḥābah (Companions).

light to whomever He willed during the life of the Prophet ﷺ. The Followers believed that the Companions ﷺ were selected by the Divine to don this honorable cloak. Thus, they coined the maxim: *All the Companions ﷺ of the Prophet ﷺ possess irrefutable integrity in matters of transmitting Islam!*[19]

THE INOCULATION OF THE CONTENT OF ISLAM FROM CONTEXTUAL INVASION

This was the historical narrative that the Followers witnessed and advanced to Muslims who succeeded them. In this narrative, it is clear that the Followers saw the Companions ﷺ as a standard that they had to follow in matters of salvation. In matters of salvational theology, the Followers witnessed the Companions ﷺ relying solely on revelation – not on their supreme intellect – at the crossroads of Muslim political history.

Along with their teachers, the Followers witnessed horrific political incidents, such as the differences between ʿAlī ﷺ and Muʿāwiyah ﷺ, and the atrocities committed by Ḥajjāj ibn Yūsuf. Despite these apparent internecine divisions, neither the Companions ﷺ nor the Followers **ever** called for religious reform or revision. They both understood that if Islamic creed and practice was to pass the test of (human) history, it needed to weather this "perfect storm". They showed tremendous religious resilience and theological integrity by refusing to allow the *context* of their history to re-shape the *content* of their Islamic identity. They chose not to question the religious integrity of their predecessors, even though they witnessed first-hand some of their political indiscretions. They

19 The maxim in its original reads as: *"al-ṣaḥābatu kulluhum ʿudūl."*

did not see any of the Companions ﷺ calling for a new religion because of political injustice and instability. It was as if they had ascertained that Islamic creed and practice was being inoculated through these horrific events against alien insinuations.[20]

In other words, the pathogen of political unrest served as a vaccine against religious mutations (*bidaʿ*). The mainstream Muslim body developed antibodies against these pathogens and did not require any invasive procedure to save it from dying. The Ummah survived and thrived by following the understanding of the Companions ﷺ in matters of religion, not by initiating a new method (*bidʿah*) for interpreting revelation. They did not find any precedent – neither in Divine Revelation nor in the acts of the Companions ﷺ – that warranted a shift in their theological paradigm. After believing in the existence of One God, the finality of the Prophethood of Muhammad ﷺ, and the resurrection on the Day of Judgment, Sunnis also believe in and follow the interpretation and legal reasoning of the Companions ﷺ in matters of religious creed and practice. This is the origin of Sunni orthodoxy and epistemology.

THE SUNNI SYSTEM OF ACQUIRING KNOWLEDGE (EPISTEMOLOGY)

By following the course chartered by their teachers, the Followers drew the map of Sunni Islam and afforded the Muslim community a path to salvation based on Allah's Grace (*Faḍl*). Sunnis believe that

---

20 Similar to the concept of vaccination, the early communities were given the pathogen of political disunity as a means to develop the theological and spiritual immune system of the Ummah. Thus, it is interesting to note that soon after these normative maxims came into being, widespread trials and tribulations afflicted these communities. It is easy for us to say this, having been removed from that history. Hindsight, of course, is 20/20.

Muslims are obligated to act according to the Sunnah (living tradition) of the Prophet Muhammad ﷺ, as conveyed by the Companions ﷺ. They should then entrust their results and judgment to Allah. Rulers and their advisers do indeed make political mistakes, but such mistakes do not take them outside of the fold of Islam. They are to be advised – like all other human beings – if they deviate from acceptable behavior. If rulers and their governments behave unjustly, they should be reprimanded (through methods prescribed by the Sunnah) but not necessarily ousted. Procuring political leadership is not the primary goal of religious leaders.[21] People other than Prophets are fallible and are therefore not automatically consigned to Hell if they fall prey to human imperfection. Likewise, God is not held hostage to any contingency in Sunni Islam. He is left "free" to do whatever He chooses. Man is to be questioned for what he does, but God is never to be questioned for what He does! This is a Sunni maxim.[22]

The Followers were nuanced – with great precision – on the rules of salvation and political engagement. They saw very clearly that the primary goal of Islam is to afford salvation for all human beings – with or without political law and order in place. No doubt, they believed that working towards building and helping a government where political law and order can regulate mundane affairs through the application of Islamic law in its entirety is an Islamic ideal and objective. However, they stopped short of demanding the establishment of a perfect political system as a requirement for an individual's salvation. For Sunni Muslims,

21 This is the theory of Sunni political quietism.

22 This is from *Surah al-Anbiyā'*, 23: "*He is not questioned about what He does, but they will be questioned.*"

salvation is independent of (and free from) political prerequisites. Those who are in positions of authority carry an extra burden to act responsibly according to the Sunnah, but they are not required to change any religious creed or practice while doing so. Thus, the primary **content** of Islamic creed and practice is not at all influenced by any historical **context.**[23]

A BRIEF HISTORY OF OPPOSING VIEWS TO THE FOLLOWERS' EPISTEMOLOGY

Groups such as the Kharijites did not follow the Companions ﷺ, instead deciding to navigate their ideology based on a call for both political and theological reform.[24] The Kharijites condemned both ʿAlī ﷺ and Muʿāwiyah ﷺ for not following the letter of the law. They saw both of them (and their predecessor ʿUthmān ﷺ) as incompetent in the eyes of Allah. By dishonoring these Companions ﷺ, they ventured against mainstream Sunni Islam and developed a new theological theory of Islam. Their theory was premised on the idea that Divine Justice must prevail in all domains of human life, and that consequently God cannot forgive anyone who commits major sins. They maintained the doctrine of ecclesiastical sentencing and denounced many Muslims as *murtaddūn* (apostates). The Kharijites held God hostage to His Justice.

Ironically, they enforced their zealous theological mission with ruthless violence; in the name of political and theological reform, they committed the very same atrocities they sought to reform. The

23 Of course, a more detailed discussion is needed in order to explain Sunni legal philosophy – but it is not the focus of this work.

24 Sunni scholars refrained from adorning these Muslims with the honorific title of Tābiʿūn (the Followers).

Kharijites drew the map for the Muʿtazilites, who were to follow their lead.

Unlike their Sunni counterparts – who entrusted the meaning of the Divine Word to the Prophet ﷺ, who received It – the Muʿtazilites embarked on a perilous journey by giving human reason primary control in deciding the meaning of the Divine Word. But to give human reason a free hand in religious affairs was to open the floodgates of theological anarchy. The Muʿtazilites were cognizant of this imminent predicament and sought to "fence in" human reason by borrowing and applying Hellenic paradigms and theories to their discourse.[25] In their quest to become Muslim Puritans, their fate followed the norm of their Hellenic role models. The alien constructs they used to substantiate revelation through reason did not withstand the test of time. Like the Greek philosophers before them, the Muʿtazilites splintered into sub-sects themselves.[26] Despite claiming the title of being champions of unity in Divinity (*tawhīd*), they became the notorious characters of Muslim sectarianism.[27]

For example, some of them held that the world is eternal as matter and is indestructible. Others chained human beings to predestination and maintained the doctrine of fatalism as being true to Islamic theory. Consequently, they served everyone redemption without a savior. Several others held human beings accountable for all their actions – bar none. In brief, the Muʿtazilites hijacked theological discourse by giving human reason authority *over* revelation itself. This ostensibly liberal approach to understanding revelation came with a heavy political price. Like their Kharijite predecessors, the

---

25 The early Abbasid rulers took on a monumental task to translate Greek philosophy into Arabic.

26 Such as the *Qadariyyah, Jabariyyah*, and *Murji'ah*.

27 Ironically, the Mu'tazilites prided themselves in being known as "The People of Divine Justice and Unity (*ahl al- 'adl wa al-tawḥīd*)".

Muʿtazilites were guilty of severe political injustices in the name of theological reform. They enacted one of the earliest and most vicious religious inquisitions in history, known as the Miḥnah (the ordeal).[28] Religious inquisitions are not designed to be peaceful, and the Miḥnah was no exception! These liberal interpreters of the Divine Word authorized conventional methods of religious persecution. Again, in the name of theological reform, we see a sect endorsing what it presumably set out to eradicate: political injustice. Although this discussion requires much more attention, it is not the focus of this introduction. This amount of background information should suffice for what I have to say from here on.

CONCLUSION TO OPENING REMARKS: THE COMPONENTS OF SUNNI EPISTEMOLOGY

As we have discussed, the debate among early Muslims was primarily in whether or not revelation can be contextualized and, if so, when and how much. Sunnis used the Companions ﵃ and their understanding of revelation as the fulcrum and pinnacle of their epistemology in primary matters of religion. Other unorthodox groups used the human intellect as an authentic standard in their epistemological framework. However, both the orthodox Sunnis and the Muʿtazilites believed in the following three tools or sources of acquiring knowledge:

- Human sensory perception (*al-ḥawwās al-khams*).
- Intellectual reasoning (*ʿaql*).
- Divine Revelation (*waḥy*).

28 Sunnis believe that the Qur'an is the Divine and uncreated Word of God. The Muʿtazilites hold that the Qur'an is created. Through their influence over certain Abbasid rulers, the Muʿtazilites persecuted those who held the Sunni view.

Sunni Muslims maintained that truth and certainty are perceptible if all three tools of acquiring knowledge are used in order of their authenticity and taxonomy. Knowledge gained through sense perception is indeed valid – and even applicable – in legal (*fiqhī*) matters, such as the timings of prayers and the purity of different types of water. The five senses are fallible, so the faculty of reason can be used to correct and "patch up" what legal realism fails to consider and determine. However, reason also is handicapped in regard to knowing or ascertaining truths and values that come before and after time. Where universal maxims require an acontextual platform from which to operate, *waḥy* comes to the rescue of the five senses and reason. *Waḥy* affords us knowledge of acontextual maxims and realities outside of time and space. Salvation is therefore based on employing all three tools of knowledge: by confirming what *waḥy* says, by using reason to negotiate how and where to apply *waḥy*, and by outwardly manifesting Islam through the physical body and five senses.

For the purpose of introducing this particular work by Imam al-Ghazālī, we need to clarify his position on orthodox Sunni doctrine. Before we do this, we must ask the following questions:

1. How much did Muslim thinkers like al-Ghazālī appreciate an empiricist approach to knowledge?

According to Muslim epistemology, the five senses are part of the three tools of knowledge: sensory perception (*al-ḥawwās al-khams*), intellectual reasoning (*ʿaql*), and Divine Revelation (*waḥy*). Muslim scholars like al-Ghazālī certainly appreciated and utilized sensory perception and its associated branches of knowledge (e.g., empirical science) in matters of faith. For example, the task of calculating ritual prayer times and determining the direction of prayer (*qiblah*) served as an impetus for Muslim scholars to develop, refine, and apply sciences such as trigonometry in their affairs. Al-Ghazālī strongly believed that a religious scholar should have knowledge of even mundane subject matters. He believed that when a Muslim makes a judgment on a topic without knowing the subject matter, he is hurting Islam.[29] Thus, Muslim scholars were never averse to utilizing the empirical sciences to facilitate the practice of life and religion.[30]

2. What role do Muslim scholars assign the other two sources of knowledge?

Muslim scholars viewed both *ʿaql* and *waḥy* as higher sources than the empirical sciences. While Muslims never saw empiricism as the **final** authority or goal, they saw it as a **component** towards holistic understanding. Moreover, knowledge gleaned from the empirical sciences that pertained to matters of salvation had to be vetted by these two sources. Even within these two sources, the mind is susceptible to error. Thus, man relies on Divine Revelation as the ultimate arbiter in the hierarchy of epistemology. For Muslims, if either of these two lower sources conflicts with Divine Revelation on a particular matter, they must not only put aside these two sources, but must be assured that using Divine Revelation is a duty and privilege of the highest order. Al-Ghazālī mentions in his *Munqidh* that the role of the mind is paramount, but Divine Revelation comes as a form of Allah's Mercy to correct the errors of the mind (*shifāʾ*). In other words, the mind has a role, but Divine Revelation can (and often will) trump all sources.

---

29 From *Tahāfut al-Falāsifah*: "The harm inflicted on religion by those who defend it not by its proper way is greater than [the harm caused by] those who attack it in the way proper to it. As has been said: '*A rational foe is better than an ignorant friend*.'"

30 It should again be stressed: but not to prove religion itself.

3. What degree of certainty does acquired knowledge need to offer before a Muslim scholar deems it credible?

Sunni Muslims deem credible only knowledge gained from conclusive (*qaṭʿī*) sources. In matters not directly related to salvation, using inconclusive sources (*ẓannī*) to facilitate life, such as the knowledge of science and the physical world, was not necessarily seen as a threat to salvation. For example, there was an incident during the lifetime of the Prophet ﷺ when he saw the Companions رضي الله عنهم cross-pollinate dates, so he commented passively about not performing this.[31] The following year, the dates failed to reach their usual harvest, so the Companions رضي الله عنهم reported this to the Prophet ﷺ. He remarked, "You are more knowledgeable about the affairs of your lives (*antum aʿlamu bi umūri dunyākum*)." This incident shows that the Prophet ﷺ was not sent to teach science, and that the Companions رضي الله عنهم were free to use their minds, experiences, and worldly knowledge to facilitate their daily affairs.

4. For inconclusive matters, do Muslim scholars follow a deontological paradigm, namely, one that is based on what a Muslim should do as a duty? Or do they follow a teleological paradigm that focuses on the consequences of knowledge? Or do Sunnis go with the approach of evidentialism that calls for scholars to follow evidence?

Sunni Muslims maintain that in order to follow the Sunnah in the manner demonstrated by the Companions, one must incorporate **all three** approaches. Sunni Muslims do not separate religion into parts. Indeed, this separation is viewed not only as artificial, but one

---

31 This hadith is known as the hadith of *taʿbīr al-nakhl*. In the narration from *Ṣaḥīḥ Muslim*, the Prophet ﷺ said: "*When I give directions in matters of your religion, you should follow them, and when I tell you something on my own, I am but a human being*" (*Muslim, Kitāb al-Faḍā'il*, hadith #4357).

that leads to division as well. The Qur'an states, "*Indeed, those who separate their religion and become splinter groups, you have nothing to do with them.*"[32] This verse does not necessarily refer to only socio-political divisions, but also points to the significant consequences that follow methodological distinctions.

It is quite amazing that the Muslim civilization – which was founded on irrevocable religious values, such as belief in the One God – ventured into debating subtle nuances in the noetic realm. Noetic sciences presume the use of the mind, whereas religious discourse presumes the application of pure belief. Although Sunnis did not afford noetic sciences a primary role, they were not at all afraid of engaging in a discussion about them.

The orthodoxy maintained that Divine Revelation (*waḥy*) is either the only or final arbiter in deciding whether a given source of knowledge is conclusive or inconclusive. Hence, inferences drawn from definite sources were deemed conclusive (*qaṭ ʿī*) if no other interpretation was plausible. Knowledge based on such sources formed the greater part of normative Islam. Classical Muslim scholars discussed how to map the human mind and psyche by using terms such as *qalb*, *rūḥ*, and *nafs*.

Conclusions based upon inconclusive sources – whether duty-based, results-based, or evidence-based – were deemed inconclusive (*ẓannī*). As a result, knowledge gained from these sources did not yield normative certainty. Examples of knowledge based on conclusive sources of *waḥy* include dogmas such as the existence of the Divine with all His Names and Attributes, normative tenets of Islam (e.g., *salāt*, *zakāt*, *sawm*, etc.), and details of eschatology. The orthodoxy saw no reason to question or interfere with such dogma.

---

32 *al-An ʿām*, 159.

They regarded anyone who challenged them as either innovators or heretics. Notably, the orthodoxy refrained from calling such challengers non-believers if they did not adhere to these dogmas that were based on inconclusive knowledge and which were regarded as non-essential in terms of belief (*ʿaqāʾid*). However, traditionalist scholars still maintained that following secondary sources of knowledge in terms of *practice* was necessary. This was especially true if there was supporting evidence to corroborate the idea of the source in question. Others – like the Ḥanafī scholars – required more rigorous authentication of such sources before they dared to classify the rule(s) in those sources as necessary. The Muʿtazilites flatly rejected any source – primary or secondary – as being conclusive if it did not measure up to what they deemed as sound human reason. The Muʿtazilites introduced philosophical principles to determine religious values.

IMAM AL-GHAZĀLĪ AND SUNNI EPISTEMOLOGY

As a trained traditionalist, Imam al-Ghazālī ventured to champion the epistemology of Sunni scholars by responding to the writings of Muslim philosophers. He did this by using the language of the philosophers – which was heavily drenched in Hellenic values – to facilitate communication with them. Most of his polemical writings are directed towards Muslim philosophers and not the traditionalists, as that would have been tantamount to preaching to the choir. He always held the view that Divine Revelation should be represented only as the Ummah (the early Muslims) had understood It. Like the Sunni theologians before him, al-Ghazālī appreciated the importance of all three sources of knowledge, while maintaining the primacy of Divine Revelation as the highest of these sources. But he also found usefulness in the other two lower sources of knowledge and sought to expose this utility in many of his writings. We can now appreciate his interest in various branches of knowledge that may not seem to yield beneficial knowledge to the uninitiated seeker.

Having discussed the approach of Imam al-Ghazālī to knowledge, we can use this epistemology as a lens to study his approach to human anatomy. By doing this, we have answered the question: why does a Muslim theologian and jurist have an interest in studying human anatomy in the first place?

IMAM AL-GHAZĀLĪ AND ANATOMY

Human anatomy as understood by modern-day physicians is based on empirical data. In this construct, human observation is essential. The tools of human observation are sensory perception and analytical reasoning. Human anatomy, as understood by religious traditions, includes the added tools of Divine Revelation and Scripture. Likewise, in the Muslim tradition, the human body has been a central object of observation and reflection.

The example of the human heart illustrates these levels of understanding. When a non-religious practitioner observes the human heart, empirical observations lead to certain conclusions. For example, the scientist understands and appreciates the heart as an organ that receives deoxygenated blood and pumps oxygenated blood to the rest of the human body. The scientist also understands that without the heart functioning optimally, the human being may die.

In a religious tradition such as Catholicism, the religious practitioner will look at the very same heart that a non-religious practitioner views. But he will add a certain set of subjective values to the heart, such as it being the central location where the soul resides. In Muslim sources of Revelation, the heart (*qalb*) is given a much broader role and value to include both empirical anatomy as well as spiritual anatomy.[33] Compared to his counterparts, the Muslim practitioner will therefore have a broader platform from which he will observe the same heart; this is because after understanding *waḥy*, he has applied all three levels of knowledge in this area of study.

For the purpose of discussing this treatise, *Concerning Divine Wisdom in the Creation of Man (Fī Ḥikmat Khalq al-Insān)*, we need to appreciate the honesty of Ḥujjat al-Islām Imam al-Ghazālī in applying the anatomical values of the human body as well as *waḥy* based values. That will lead the reader to this broader platform of understanding and appreciating the Creator. This contribution to our understanding of human anatomy – which uses all three sources of knowledge – distinguishes the approach of Imam al-Ghazālī to human anatomy from others, including that of other Muslim anatomists as well. We can further appreciate this contribution by briefly discussing the role played by one of the most well-known physicians in Muslim history: Abū ʿAlī al-Ḥusayn ibn Sīnā.

In the history of medicine, Ibn Sīnā has an obvious significance, as he had mastered the sciences of human anatomy at a very young age. Along with his mastery of contemporary medicine, he was also well-regarded for his philosophical acumen and academic contributions. His seminal work, *al-Qānūn*, was taught as a textbook in universities across the world for centuries and was a standard reference text for all physicians. Imam al-Ghazālī succeeded Ibn Sīnā and certainly must have read his texts. However, he differed with Ibn Sīnā, most notably in the disciplines of rational philosophy and theology, as well as their

---

33 While there are many examples of this, a well-known hadith states: "*Indeed, in the body there is a lump of flesh. If it is sound, the entire body is sound, and if it is corrupt, the whole body is corrupt; indeed, it is the heart*" (Bukhārī and Muslim).

role in the hierarchy of acquiring knowledge. This demonstrates that Muslim scholars were not averse to applying the empirical method to their studies. In fact, combining empirical knowledge, an understanding of *waḥy*, and behavior based on spiritual discipline helped scholars form a holistic worldview on diverse subjects. I believe that readers of this book should appreciate al-Ghazālī's incorporation and use of these faculties of knowledge which give him a unique vantage point to observe and comment on human anatomy.

AL-GHAZĀLĪ'S VANTAGE POINT

We need to expand on the broader platform that Imam al-Ghazālī brings to this discussion. As mentioned previously, Muslims believed that there were three tools for acquiring knowledge. The purpose of this introduction is to understand how he used the first two tools (empiricism and intellectual reasoning) to launch himself further and higher to appreciate the third tool (Divine Revelation) and ultimately understand what the Divine says about the human body. While he had mastered the religious sciences, al-Ghazālī ventured to look deeper into Divine Revelation and the authentic narration of the Prophet (*khabar al-Rasūl*) to find deeper meanings in texts which he had earlier deemed as purely legal or philosophical. This deeper insight came as a result of what he himself termed as spiritual cleansing.[34] This power of observation allowed him to use all three tools to reach a level of understanding that was greater than the sum of its parts.

An analogy may be helpful here. Man's ability to see has multiple layers. For example, on one level, physical eyesight relies on the human eye and mind to observe a chair that is present in a room.

---

34 His book *al-Munqidh* is an expression of this development.

Above this, man's intellectual eye allows him to visualize a chair in his mind without actually observing one. Finally, the spiritual eye allows one to observe truths that are invisible and unimaginable to these other types of vision, as in the case of belief in the Verse of the Chair.[35] Muslims believe in this Chair as an entity that is more real than the matter in our worldly material realm. Yet no ordinary human eye has seen it nor has any ordinary human mind been able to imagine it. Therefore, because human beings cannot directly access this third level of eyesight, knowledge of this level is only retrievable through the authentic narration of the Prophet ﷺ. Once Imam al-Ghazālī reached this third level of observation, he re-wrote his understanding of the physical and metaphysical worlds. The third level of observation is grounded in Islamic theology, which requires an authentic understanding of the narration of the Messenger ﷺ. Muslims have debated the methodology towards understanding theology from a very early time historically. The previous summary on Sunni theology helps us appreciate this level more appropriately.

This treatise is actually a small chapter within a larger work titled *Divine Wisdom in the Creation of God the Almighty and Exalted* (*al-Ḥikmatu fī Makhlūqātillāh ʿAzza wa Jall*). It is believed that this work was written in the latter period of al-Ghazālī's life, when his beliefs on epistemology had crystallized after years of study, introspection, and spiritual training. The other treatises in this work have elements of al-Ghazālī's application of this epistemology. We have chosen this particular treatise to serve as an exposé on the human body. In this work, al-Ghazālī uses his acumen to observe and describe the limbs and organs with eloquent precision that defy our standards of expression.

---

35 *Āyah al-Kursī* i.e., *Surah al-Baqarah*, 255.

In reading this text, a Muslim medical practitioner should be able to understand the full extent of his role in appreciating Divine Wisdom in his chosen field of study. Whereas Ibn Sīnā revolutionized anatomy and medicine with *al-Qānūn*, Imam al-Ghazālī provided us a greater platform from which we may observe the former's findings on the human body. Whereas *al-Qānūn* is purely a book of science, this short treatise is grounded in Islamic metaphysics and theological principles. The agenda of Imam al-Ghazālī is to train the outward eye to observe, analyze, and introspect, in order to train the inner eye to see the realities of the holistic cosmos in this world.

THE CREATION OF THE FIRST TWO HUMAN BEINGS

Allah created Ādam ﷺ from clay, and his body did not have any life until Allah breathed His Spirit (*rūḥ*) into it. Muslim epistemology maintains that the body and *rūḥ* are separate entities, with the latter being under Divine Control, given its origin.[36] Whereas the *rūḥ* is from the higher order of the cosmos, the human body is from the lower order (the Earth).[37] When Allah wished to establish the *khilāfah* (vicegerency) of Ādam ﷺ, He empowered the lowest form of matter (the Earth) to contain the highest order of matter (the *rūḥ*). This Divine Grace conferred nobility on Ādam ﷺ and his progeny. Allah inserted the *rūḥ*, the highest form of creation, into the lowest form such that the two combined to become one without any dichotomy. Hence, human accountability is on both the soul

---

36 *al-Isrā'*, 85: "*They ask you (O Muhammad) about the soul. Say, 'The soul is of the affair of my Lord, and you have not been given knowledge of it except a little.'*"

37 The *turāb* mentioned in the Qur'an is from the Earth. Every representative sort of earthly matter went into the creation of Ādam when Allah ordered the Angel ʿAzrā'īl to collect all the different types of soil; *judad* (black), *ḥumurun* (red), and so on, as per *Surah al-Fāṭir*, 27. All of the various elements of the Earth are found in the human being.

and the physical body, and Sunni Muslims maintain that both will be resurrected and held responsible before Allah.

At the time of the creation of Ādam ﷺ, the Angels expressed concern about the eligibility of his being a vicegerent, but Allah replied to them: "I know what you know not."[38] The Angels knew that anything created from earthly matter and living in the Earth (*dunyā*) was subject to violence and bloodshed, but they did not know that Allah was going to breathe His Rūḥ into this earthly matter: this is the miracle (*mu ʿjizah*) concerning our creation. The creation of Ādam ﷺ is therefore much more of a miracle than the creation of ʿĪsā (Jesus) ﷺ. Ḥawwā' was created from Ādam ﷺ as an independent human being. Both lived as husband and wife in Jannah, the perfect place of residence. Both Ādam ﷺ and his wife were created outside of the Earth, even though their essence is from it. Their creation was perfect because they were to be placed into Paradise and serve as a prototype for all human beings that were to descend from them and eventually return to this perfect place.

When Ādam ﷺ and his wife came down to this Earth, the creation of their children was to occur completely here, and therefore, the mechanism for procreation was activated in this world. This mechanism included what we know about the *rūḥ* being inserted into the fetus while in the womb. Imam al-Ghazālī not only respected the human body but made the understanding of how it comes into full being a part of his theology. There is a value to life before the *rūḥ* comes into the fetus, which is known as the animalistic state of life (*al-ḥayāh al-ḥayawāniyyah*), that lasts

38 *al-Baqarah*, 30: "*Remember when your Lord said to the Angels, 'Indeed, I will make upon the earth a successive authority.' They said, 'Will You place upon it one who causes corruption therein and sheds blood, while we declare Your praise and sanctify You?' Allah said, 'Indeed, I know that which you do not know.'*"

for the first 120 days. When the *rūḥ* enters the physical body, the fetus is now in a fully human state of life (*al-ḥayāh al-insāniyyah*).[39] Only the Angels can "surgically" insert or remove the *rūḥ*, whether in the womb or at the time of death. There is a profound subtlety at this time of insertion, as the mother does not know when the *rūḥ* enters the womb, and her life is not violated with the transfer of this powerful entity from the spiritual world (*ʿālam al-arwāḥ*) into this world. Yet at the time of death, when the soul leaves the body, there is again a transfer of this powerful entity such that the human being dies, but the physical body does not immediately disintegrate.

THE RESURRECTION OF HUMAN BEINGS

Imam al-Ghazālī disagreed vehemently with Muslim philosophers, who argued that the physical body will not be resurrected. Sunnis premised their doctrine of physical resurrection on taking matters of dogma as they are stated in Divine Revelation. The question, then, is how our bodies will function in the world of the Hereafter. We have seen how the understanding or perception of the word "chair" differs according to the tools used to observe it. According to Sunni belief, the perceptible ability of senses is heightened as soon as the human being departs this world. This is in contrast to Prophets, who are given the ability to witness realities of the next world in this world itself.

For example, we believe that a person in the grave will perceive and understand both punishment and pleasure.[40] In his works, Imam

---

39 In some of his writings, al-Ghazālī was of the opinion that there is a difference between *al-ḥayāh al-ḥayawāniyyah* and *al-insāniyyah*. The Muʿtazilites, including Ibn Sīnā, accepted this outside of this concept.

40 For example, regarding punishment: Ibn ʿAbbās reported: "The Messenger of Allah ﷺ

al-Ghazālī likens this to the perception of a person who dreams and experiences pain and pleasure in that dream. Although the dreamer has very real experiences during the dream, it becomes a secondary memory that remains (often only partially) in his mind when he awakens, and does not exist in his everyday life. For the person in the grave, this perception is much higher, and the resulting pain and pleasure is much more real.

Imam al-Ghazālī argues that the realities of the afterlife are to be observed through the lens of the dreamer. Like the dreamer, the one in the grave may feel pain and pleasure without the physical body being part of this experience. After this time in the world of graves (*barzakh*), the human being will be resurrected on the Day of Judgment. Sunni Muslims believe that the physical body will conform to the state of the spirit (*rūḥ*), allowing the senses to be heightened. If a person has lived a moral life, his body will appear in a good form. If a person has lived an evil life, his body will likewise appear in this form. However, both will possess heightened senses.

In the realm of the Hereafter, time will be expanded.[41] This

---

passed by two graves and said, 'Both of them are being punished, but not by a major sin. As for this man, he did not prevent his urine from soiling him. And as for this man, he would spread gossip.' The Prophet then asked for a green leaf from a date-palm. The Prophet split it into two pieces and planted one on each grave, then he said, 'It is hoped that their punishment may be abated until those two leaves become dry.'" (*Ṣāḥīh al-Bukhārī* 5705, *Ṣāḥīh Muslim* 292). As for reward, one example is as follows: Abū Hurayrah reported: The Messenger of Allah ﷺ said, "When the deceased is buried in his grave, two black and blue Angels come to him. One of them is called Munkar and the other is called Nakīr. They will both say, 'What do you say about this man?' He will say what he said before, 'He is the servant of Allah and His Messenger. I bear witness there is no God but Allah, and Muhammad is His servant and His Messenger.' They will both say, 'We knew you would say this!' Then his grave is expanded by seventy cubits on each side. It will be illuminated for him, and then it will be said to him: 'Sleep!' He will say, 'May I return to my family to tell them?' They (Munkar and Nakīr) will both say: 'Sleep like a bridegroom, whom none awakens but the most beloved of his family, until Allah resurrects him from his place of rest...'" (*Sunan al-Tirmidhī*, 1071).

41 For example, the Qur'an indicates that a day in supra-earthly realms is much longer

suggests that the abilities of human beings will be uniquely optimized to perceive the realities of both that Day and that realm. The realities of this Day and beyond are to be understood only through authentic narrations from the Prophet ﷺ. In this realm, the limited abilities of the human being in the physical world become negligible as the *rūḥ* will dictate how much a human being is able to perceive and grasp. Despite their disagreement on the resurrection of the physical body, Muslim philosophers agreed with this idea regarding the soul's heightened sense of perception. An ordinary human being's imagination is not able to perceive any of the realities of this realm, as is commonly agreed upon in Sunni doctrine. Once we have understood the Sunni understanding of physical resurrection, we can direct our attention to how each of these physical senses will be heightened in the next world.

THE FIVE SENSES IN THE HEREAFTER

Through the Qur'an and Sunnah, Sunni Muslims maintain that all five senses will be enhanced with supranatural abilities in the Hereafter. We will comment on two of these senses here. The eyes will be given an ability to see the Face of Allah, and the ears will be given an ability to hear Allah's Recitation of the Qur'an. This is a remarkable occurrence that is exclusive to this realm because, where previously (in this world) there was no relationship (*nisbah*) between the Creator and creation in certain faculties, Allah will create a type of relationship that cannot be rationally explained or understood. Sunni Muslims therefore maintain that despite our

---

than our perception in this world, ranging from 1,000 years (*al-Ḥajj*, 47) to 50,000 years (*al-Ma'ārij*, 4).

inability to fully comprehend these statements, it is a reality beyond our known reality, and can only be understood in the Hereafter.

From this theology of heightened senses, we see how al-Ghazālī draws a conclusion about pleasure in Paradise. Based on a Prophetic tradition, he maintains that the height of ecstasy and pleasure is to look at the Face of Allah. Therefore, his interest in this subject is profoundly simple: how do you develop and train your eyes so that you are able to look at Him and derive the greatest of pleasures? It is as if he is saying: *if you are fascinated with the makhlūq (creation), then you should be even more fascinated with the Khāliq (Creator).* When the Muslim practitioner reads anatomy with this lens, a feeling of gratitude and appreciation for the Creator results in the verbal phrase and internal feeling of *alḥamdulillāh* (all praise is due to God). For al-Ghazālī, his focus is understanding and appreciating the breadth of meaning found in this term. Embracing *alḥamdulillāh* – verbally, intellectually, spiritually, and philosophically – gives us an appreciation for all the wonderful praiseworthy acts of Allah, which include the act of bestowing upon His creation the gift of seeing Him and listening to His Recitation from the Noble Qur'an. The medical professional may wish to consider this vantage point when reading this anatomical treatise.

We came across a fascinating treatise of al-Ghazālī on the Divine Wisdom contained within a detailed survey of human anatomy. This work of Abū Ḥāmid is one that hopefully captures the imagination of a Muslim physician – even at first glance. His ability to observe some of the minute details of the human body through empirical methods is quite remarkable. There is no doubt that he must have read contemporary manuals on human anatomy and that he must have consulted with the physicians of his time. But the fact that he was able to observe certain utilitarian marvels of human anatomy with far more pristine clarity than any medical empiricists of any era is a testimony to his command of rational and metaphysical sciences. For example, every intelligent human being knows that the hair on our heads is significantly different from the hair on other parts of our body. But rarely has any physician or philosopher expressed this obvious fact more eloquently than al-Ghazālī does:

> *So [observe] the...purpose of eyelashes [which are] for the beautification of the eye and face, He made its hair to be a specific length and measure, not containing more (in length or thickness) such that it would harm the eye, nor being incomplete in length (which would render it useless for the purposes of beautification and protection). He created saltiness in the conjunctiva of the eye in order to destroy whatever may fall into it. He made the two sides of the eye as two small depressions [in comparison to the level] from the center [of the eye], so that anything that fell into the eye would be directed towards these outer margins. He made the two eyebrows as a beautification of the face and a covering (protection) for the eyes, and He made their hair to resemble eyelashes in its lacking the ability to grow long and become misshapen. Yet He made the hair of the scalp and beard (which is also hair like eyebrows) capable of growing and being shortened. [Appreciate] the beauty that was intended in these two [eyebrows] without any distortion therein.*

Imam al-Ghazālī wants the reader to move beyond simply appreciating the anatomical wisdom of how the human being is created here in this world, but to appreciate how this will be even more sophisticated in the next world. This agenda is in line with the purpose of his magnum opus (*Iḥyā' 'Ulūm al-Dīn*), which is to purify the human body through character reformation and spiritual cleansing such that the human being may transition smoothly into the next world. If there is filth or defects on these bodily organs, they will not be prepared to experience the bliss of the next world. Imam al-Ghazālī wants us to appreciate our anatomy **here** so that it does not jeopardize our experiencing of bliss over **there**. Through character reformation and spiritual discipline, Imam al-Ghazālī provides us with a path to climb the tree of *tawḥīd* and taste the fruits at the top – while the philosophers and scientists leave us stranded on a limb.

Concerning Divine Wisdom in the Creation of Man

Ḥujjat al-Islām
Abū Ḥāmid al-Ghazālī ﵀ (1058-1111)

Translation by Dr. Kamran M. Riaz
Commentary by Sheikh Mohammed Amin Kholwadia
and Dr. Kamran M. Riaz[i]

THE MOST EXALTED SAID: "*Indeed We created man from an extract of clay...*" until the end of what He described in the verse.[ii] Know – and may God give you guidance – that God, Mighty and Majestic is He, when (the idea of) Creation of all things came to be according to His Knowledge, He dispersed them in this world, and He commissioned for them trials and examinations therein. He (may He be Glorified!) created them with some of them procreating others. Then, He created the male and female, and placed in their hearts love and exigencies (towards each other), until they were incapable of patience and devoid of any stratagem in refraining from carnal desires. This primordial drive for closeness drove them in their naturally created states towards intimacy.

بَابٌ فِي حِكْمَةِ خَلْقِ الْإِنْسَانِ

حجة الإسلام
أبو حامد الغزالي رحمه الله
ترجمة كامران محمد رياض
تعليق الشيخ محمد أمين خولواديا و كامران محمد رياض

قَالَ تَعَالَى: {وَلَقَدْ خَلَقْنَا الْإِنْسَانَ مِنْ سُلَالَةٍ مِّنْ طِينٍ} [المؤمنون:١٢]، إِلَى آخِرِ مَا وَصَفَهُ سُبْحَانَهُ. اِعْلَمْ - وَفَّقَكَ اللهُ تَعالى - أَنَّ اللهَ عَزَّ وَجَلَّ لَمَّا سَبَقَ فِي عِلْمِهِ خَلْقُ الْخَلْقِ وَبَثُّهُم فِي هَذِهِ الدَّارِ، وَتَكْلِيفُهُمْ فِيْهَا لِلْبَلْوَى وَالِاخْتِبَارِ، خَلَقَهُمْ سُبْحَانَهُ مُتَنَاسِلِيْنَ بَعْضُهُمْ مِنْ بَعْضٍ، فَخَلَقَ سُبْحَانَهُ الذَّكَرَ وَالْأُنْثَى، وَأَلْقَى فِي قُلُوبِهِمُ الْمَحَبَّةَ وَالدَّوَاعِي، حَتَّى عَجَزُوا عَنِ الصَّبْرِ، وَعَدِمُوا الْحِيْلَةَ فِي اجْتِنَابِ الشَّهْوَةِ، فَسَاقَتْهُم الشَّهْوَةُ الْمَفْطُوْرَةُ فِي خَلْقِهِمْ إِلَى الِاجْتِمَاعِ.

He made desirous thoughts in humanity (for one another) become capable of stimulating a specific organ for the depositing of water into a firmly-established place wherein life is created.[iii] Sperm gathers in this organ, drawing from the essence of the rest of the body, and exits as water, bursting and gushing forth from between the backbone and the ribs with a specifically ordained movement.[iv] The zygote is carried through the process of splitting and cleaving[v] (after the union of the sperm and egg) from one hidden and inner location to another.[vi] So it (the zygote) remains, in spite of its locomotion, according to its true form (to facilitate the development of life).[vii] Even though it (sperm) is a despised water, the lowest of all things – when it touches (something), it corrupts and changes the (purified) nature of things – it is still a water, all of it containing equally mixed parts (of all genetic material needed to create life), which are unchanged in any condition.[viii]

Then He (may He be Glorified) created both male and female from it, after its journey from sperm to a *ʿalaqah* (clot), to a *muḍghah* (chewed lump of flesh), until it had bones; then He covered it with flesh, strengthened it with nerves and sinews, and weaved blood vessels (into the midst of these).[ix] He created the limbs, assembled them, and made the head round in shape (upon them). (In the head) He opened forth (organs of) hearing, sight, smell, taste, and other orifices.

Then He made the eye for the purpose of sight. From its wonders is that the secret of its existence is to reveal things – and the secret of this (particular) matter is incapable of being described.[x,xi] He constructed it having seven layers, and each layer has a quality and specified shape.[xii] If the eye were devoid of any layer or possessed any excess (in layers or thickness), it would be rendered incapable of sight.[xiii] So observe the shape of the eyelids that protects the eye, and what (magnificence) He created in their ability for swift movement

وَجَعَلَ الْفِكْرَةَ تُحَرِّكُ عُضْوًا مَخْصُوْصًا بِهِ إِلَى إِيدَاعِ الْمَاءِ فِي الْقَرَارِ الْمَكِيْنِ الَّذِي يُخْلَقُ فِيْهِ الْجَنِيْنُ، فَاجْتَمَعَتْ فِيْهِ النُّطْفَةُ مِنْ سَائِرِ الْبَدَنِ، وَخَرَجَتْ مَاءً دَافِقًا مُنْدَفِعًا مِنْ بَيْنِ الصُّلْبِ وَالتَّرَائِبِ بِحَرَكَةٍ مَخْصُوْصَةٍ، فَانْتَقَلَتْ بِسَبَبِ الإِيلاجِ مِنْ بَاطِنٍ إِلَى بَاطِنٍ، فَكَانَتْ مَعَ انْتِقَالِهَا عَلَى أَصْلِهَا؛ لِأَنَّهَا مَاءٌ مَهِيْنٌ أَدْنَى شَيْءٍ يُبَاشِرُهَا يُفْسِدُهَا وَيُغَيِّرُ مِزَاجَهَا، فَهِيَ مَاءٌ يَخْتَلِطُ جَمِيْعُهُ، مُسْتَوِيَةٌ أَجْزَاؤُهُ، لَا تَفَاوُتَ فِيْهَا بِحَالٍ.

فَخَلَقَ سُبْحَانَهُ مِنْهُ الذَّكَرَ وَالْأُنْثَى بَعْدَ نَقْلِهَا مِنَ النَّطْفَةِ إِلَى الْعَلَقَةِ إِلَى الْمُضْغَةِ إِلَى الْعِظَامِ، ثُمَّ كَسَاهَا اللَّحْمَ، وَشَدَّهَا بِالْأَعْصَابِ وَالْأَوْتَارِ، وَنَسَجَهَا بِالْعُرُوْقِ، وَخَلَقَ الْأَعْضَاءَ وَرَكَّبَهَا فَدَوَّرَ سُبْحَانَهُ الرَّأْسَ، وَشَقَّ فِيْهَا السَّمْعَ وَالْبَصَرَ وَالْأَنْفَ وَالْفَمَ وَسَائِرَ الْمَنَافِذِ.

فَجَعَلَ الْعَيْنَ لِلْبَصَرِ، وَمِنَ الْعَجَائِبِ سِرُّ كَوْنِهَا مُبْصِرَةً لِلْأَشْيَاءِ، وَهُوَ أَمْرٌ يُعْجِزُ عَنْ شَرْحِ سِرِّهِ، وَرَكَّبَهَا مِنْ سَبْعِ طَبَقَاتٍ، لِكُلِّ طَبَقَةٍ صِفَةٌ وَهَيْئَةٌ مَخْصُوصَةٌ بِهَا، فَلَوْ فُقِدَتْ طَبَقَةٌ مِنْهَا أَوْ زَالَتْ لَتَعَطَّلَتْ عَنِ الْإِبْصَارِ، وَانْظُرْ إِلَى هَيْئَةِ الْأَشْفَارِ الَّتِي تُحِيطُ بِهَا وَمَا خُلِقَ فِيْهَا مِنْ سُرْعَةِ الْحَرَكَةِ؛ لِتَقِيَ الْعَيْنَ مِمَّا يَصِلُ إِلَيْهَا مِمَّا يُؤْذِيْهَا مِنْ غُبَارٍ

for the purpose of protecting the eye from whatever may reach it and cause harm to it, like dust and other such things. So (observe) the eyelids, which were created with the status and function like a door: they can be opened when there is a need (for vision) and they can be closed at other times (for sleep and to protect the eye from desiccation and physical damage). Because the purpose of eyelashes was for the beautification of the eye and face, He made its hair to be a specific length and measure, not containing more (in length or thickness) such that it would harm the eye, nor being incomplete in length (which would render it useless for the purposes of beautification and protection).[xiv] He created saltiness in the conjunctiva of the eye in order to destroy whatever may fall into it.[xv] He made the two sides of the eye as two small depressions (in comparison to the level) from the center (of the eye), so that anything that fell into the eye would be directed towards these outer margins.[xvi] He made the two eyebrows as a beautification of the face and a covering (protection) for the eyes, and He made this hair to resemble eyelashes in lacking the ability to grow long and become misshapen. Yet He made the hair of the scalp and beard (which is also hair, like eyebrows) capable of growing and being shortened.[xvii] (Appreciate) the beauty that was intended in these two (eyebrows) without any distortion therein.

Then look at the mouth and tongue and see what Divine Wisdom is contained therein.[xviii] He made the two lips as a curtain for the mouth, as if they were a door that is closed but (one that) opens when the time for such a need arrives;[xix] and the lips are (also) a covering over the gums and teeth, with the added benefit of imparting beauty (to the face). Had it not been for these (two lips), the entire creation (of the face) would be misshapen and ugly. The lips are also helpers for (the purpose of) speech. The tongue is for rational speech and the articulation of what is in man's inner self. It is also for the turning of

وَغَيْرِهِ، فَكَانَتِ الْأَشْفَارُ بِمَنْزِلَةِ بَابٍ يُفْتَحُ وَقْتَ الْحَاجَةِ، وَيُغْلَقُ فِي غَيْرِ وَقْتٍ، وَلَمَّا كَانَ الْمَقْصُودُ مِنَ الْأَشْفَارِ جَمَالَ الْعَيْنِ وَالْوَجْهِ جَعَلَ شَعْرَهَا عَلَى قَدْرٍ لَا يَزِيدُ زِيَادَةً تَضُرُّ بِالْعَيْنِ، وَلَا تَنْقُصُ نَقْصًا يَضُرُّ بِهَا، وَخَلَقَ فِي مَائِهَا مُلُوحَةً؛ لِتَقْطِيعِ مَا يَقَعُ فِيهَا،

وَجَعَلَ طَرَفَيْهِمَا مُنْخَفِضَيْنِ عَنْ وَسَطِهِمَا قَلِيلاً؛ لِيَنْصَرِفَ مَا يَقَعُ فِي الْعَيْنِ لِأَحَدِ الْجَانِبَيْنِ، وَجَعَلَ الْحَاجِبَيْنِ جَمَالاً لِلْوَجْهِ وَسَتْرًا لِلْعَيْنَيْنِ، وَشَعْرُهُمَا يُشْبِهُ الْأَهْدَابَ فِي عَدَمِ الزِّيَادَةِ الْمُشَوِّهَةِ،

وَجَعَلَ شَعْرَ الرَّأْسِ وَاللِّحْيَةِ قَابِلاً لِلزِّيَادَةِ وَالنَّقْصِ، فَيُفْعَلُ فِيهِمَا مَا يُقْصَدُ بِهِ الْجَمَالُ مِنْ غَيْرِ تَشْوِيهٍ.

ثُمَّ انْظُرْ إِلَى الْفَمِ وَاللِّسَانِ وَمَا فِي ذَلِكَ مِنَ الْحِكَمِ، فَجَعَلَ الشَّفَتَيْنِ سَتْرًا لِلْفَمِ كَأَنَّهَا بَابٌ يُغْلَقُ وَقْتَ ارْتِفَاعِ الْحَاجَةِ إِلَى فَتْحِهِ، وَهُوَ سَتْرٌ عَلَى اللِّثَةِ وَالْأَسْنَانِ مُفِيدٌ لِلْجَمَالِ، فَلَوْلَاهُمَا لَتَشَوَّهَتِ الْخِلْقُ، وَهُمَا مُعِينَانِ عَلَى الْكَلَامِ، وَاللِّسَانَ لِلنُّطْقِ وَالتَّعْبِيرِ عَمَّا فِي ضَمِيرِ الْإِنْسَانِ، وَتَقْلِيبِ الطَّعَامِ وَإِلْقَائِهِ تَحْتَ الْأَضْرَاسِ؛ حَتَّى يَسْتَحْكِمَ مَضْغُهُ، وَيَسْهُلَ ابْتِلَاعُهُ، ثُمَّ جَعَلَ الْأَسْنَانَ أَعْدَادًا مُتَفَرِّقَةً وَلَمْ تَكُنْ عَظْمًا وَاحِدًا، فَإِنْ أَصَابَ بَعْضَهَا ثَلْمٌ انْتَفَعَ بِالْبَاقِي، وَجَمَعَ

food and delivering it under the teeth, mollifying it until it is fit for chewing and becomes easy for swallowing. Then He made the teeth into different numbers, such that they are not (like) a single bone. If one of them were afflicted with a deficiency, there would still be benefit from (being able to use) the others. He combined (the faculties of) both usefulness and beauty into teeth. He made each tooth have (a tooth) that is in opposition to it and extended each one so that it met with the one below it (and thus allowed human beings to chew food properly and with ease). He made them sturdy, yet unlike the bones of the body, due to man's need for continuous usage (of the teeth).[xx] There is both size and prodigality in the molars, due to the need of chewing food, for indeed chewing is the first digestion.[xxi] The incisors and canines were made specifically for the cutting of food, yet also as a (source of) beautification for the mouth, so He strengthened their roots and sharpened their heads.[xxii] He whitened the colour of teeth, which are surrounded by the redness of gums (appearing beautiful with this contrast of colours), and made them possess symmetrical tops and a proportionate order (of arrangement), like a well-arranged row of pearls.

Then look at how He created a contained moisture (through saliva) in the mouth that is isolated and does not appear until there is a need for it.[xxiii] If it (thin saliva) was present and flowed before there was a need for it, it would be a disfigurement. It was (therefore) made to provide wetness to whatever food was to be chewed, until it became easy to swallow without any hardship or pain.[xxiv] This extra wetness, which was created specifically for the moistening of food, is also not present when one is not eating. Yet there remains enough wetness to moisten the uvula and throat for the purpose of forming speech, such that these organs do not become dry – for if they became dry, it would destroy man.[xxv,xxvi,xxvii]

Then look at the mercy of God and His Kindness: how He

فِيْهَا بَيْنَ النَّفْعِ وَالْجَمَالِ، وَجَعَلَ مَا كَانَ مِنْهَا مَعْكُوْسًا زَائِدَ الشِّعَبِ؛ حَتَّى تَطُوْلَ مُدَّتُهُ مَعَ الصَّفِّ الَّذِي تَحْتَهُ، وَجَعَلَهَا صَلْبَةً لَيْسَتْ كَعِظَامِ الْبَدَنِ؛ لِدُعَاءِ الْحَاجَةِ إِلَيْهَا عَلَى الدَّوَامِ، وَفِي الْأَضْرَاسِ كِبَرٌ وَتَسْرِيفٌ؛ لِأَجْلِ الْحَاجَةِ إِلَى دَرْسِ الْغِذَاءِ، فَإِنَّ الْمَضْغَ هُوَ الْهَضْمُ الْأَوَّلُ، وَجُعِلَتِ الثَّنَايَا وَالْأَنْيَابُ لِتَقْطِيْعِ الطَّعَامِ وَجَمَالاً لِلْفَمِ، فَأَحْكَمَ أُصُوْلَهَا، وَحَدَّدَ ضُرُوْسَهَا، وَبَيَّضَ لَوْنَهَا مَعَ حُمْرَةِ مَا حَوْلَهَا، مُتَسَاوِيَةَ الرُّؤُوْسِ مُتَنَاسِبَةَ التَّرْكِيْبِ، كَأَنَّهَا الدُّرُّ الْمَنْظُوْمُ.

ثُمَّ انْظُرْ كَيْفَ خَلَقَ فِي الْفَمِ نَدَاوَةً مَحْبُوْسَةً لَا تَظْهَرُ إِلَّا فِي وَقْتِ الْحَاجَةِ إِلَيْهَا؛ فَلَوْ ظَهَرَتْ وَسَالَتْ قَبْلَ ذَلِكَ لَكَانَ تَشْوِيْهاً لِلْإِنْسَانِ، فَجُعِلَتْ لِيُبَلَّ بِهَا مَا يُمْضَغُ مِنْ الطَّعَامِ حَتَّى يَسْهُلَ تَسْوِيْغُهُ مِنْ غَيْرِ عَنَتٍ وَلَا أَلَمٍ. فَإِذَا فُقِدَ الْأَكْلُ عُدِمَتْ تِلْكَ النَّدَاوَةُ الزَّائِدَةُ الَّتِي خُلِقَتْ لِلتَّرْطِيْبِ، وَبَقِيَ مِنْهَا مَا يَبُلُّ اللَّهَوَاتِ وَالْحَلْقَ لِتَصْوِيرِ الْكَلَامِ، وَلِئَلَّا يَجِفَّ، فَإِنَّ جَفَافَهُ مُهْلِكٌ لِلْإِنْسَانِ.

ثُمَّ انْظُرْ إِلَى رَحْمَةِ اللهِ وَلُطْفِهِ؛ إِذْ جَعَلَ لِلْآكِلِ لَذَّةَ الْأَكْلِ، فَجَعَلَ الذَّوْقَ فِي اللِّسَانِ وَغَيْرِهِ مِنْ أَجْزَاءِ الْفَمِ؛ لِيَعْرِفَ بِالذَّوْقِ مَا يُوَافِقُهُ وَيُلَائِمُهُ مِنَ الْمَلْذُوْذِ، فَيَجِدَ فِي ذَلِكَ رَاحَةً فِي الطَّعَامِ وَالشَّرَابِ إِذَا

made the ability to taste food as a part of eating, and how He made (the ability to) taste in the tongue and other parts of the mouth.[xxviii] Through (the faculty of) taste, he may know what agrees with him and what things are compatible, so that he may find comfort in food and drink when he has a need for them, and he may refrain from things that do not agree with him.[xxix,xxx] Through taste, he also comes to know the extent of hotness and coldness in the foods that reach him.[xxxi]

Then God the Exalted pierced (into the human being) the faculty of hearing.[xxxii] He placed a bitter wetness in the ears to protect them from the harm of insects, such that it kills most of the vermin that enter into the ear.[xxxiii,xxxiv] He protected the ear by the auricle, which also gathers all sound (waves) and directs it to the auditory meatus. He made in it a heightened sensitivity to be able to perceive any pests and other such things that may try to reach it.[xxxv] He (also) made many internal curvatures within it to amplify the sound (in it). In order to lengthen the (distance of) movement of that which crawls (invades) into it (i.e., an insect or other harmful thing), He made this path long, such that it may awaken the person from sleep and affect him.[xxxvi]

Then look at (man's ability to) perceive odorous things by means of the entrance of air (through the nostrils); indeed, this is a secret which none understands its true reality except for the Creator (may He be Glorified) to a degree other than (what has been mentioned here).[xxxvii,xxxviii,xxxix] Then look at how He raised the nose in the middle of the face, and made man's appearance beautiful (by placing it in this location as opposed to elsewhere on the face). He opened the two nostrils, and He placed in it a sense capable of smelling so that man may have proof through his faculty of olfaction as to the smell of his food and drink, and that he may find pleasure in the smells of perfume,[xl] guard himself against foul odors, and be able to inhale

دَعَتْ إِلَيْهِ حَاجَةٌ إِلَى تَنَاوُلِهِ، وَلِيَجْتَنِبَ الشَّيْءَ الَّذِي لَا يُوَافِقُهُ، وَيَعْرِفَ بِذَلِكَ حَدَّ مَا تَصِلُ الْأَشْيَاءُ إِلَيْهِ فِي الْحَرَارَةِ وَالْبُرُوْدَةِ.

ثُمَّ إِنَّ اللهَ تَعَالَى شَقَّ السَّمْعَ وَأَوْدَعَهُ رُطُوبَةً مُرَّةً يُحْفَظُ بِهَا السَّمْعُ مِنْ ضَرَرِ الدُّوْدِ، وَيُقْتَلُ أَكْثَرُ الْهَوَامِّ الَّذِيْنَ يَلِجُوْنَ السَّمْعَ، وَحَفِظَ الْأُذُنَ بِصَدَفَةٍ لِتَجْمَعَ الصَّوْتَ فَتَرُدَّهُ إِلَى صِمَاخِهَا، وَجَعَلَ فِيْهَا زِيَادَةَ حِسٍّ لِتَحُسَّ بِمَا يَصِلُ إِلَيْهَا مِمَّا يُؤْذِيْهَا مِنْ هَوَامَّ وَغَيْرِهَا، وَجَعَلَ فِيْهَا تَعْوِيْجَاتٍ لِيَطَّرِدَ فِيْهَا الصَّوْتُ، وَلِتَكْثُرَ حَرَكَةُ مَا يَدِبُّ فِيْهَا وَيَطُوْلَ طَرِيْقُهُ فَيَنْتَبِهَ فَيَتَأَثَّرَ وَيَنْتَبِهَ صَاحِبُهَا مِنَ النَّوْمِ.

ثُمَّ انْظُرْ إِلَى إِدْرَاكِهِ الْمَشْمُوْمَاتِ بِوَاسِطَةِ وُلُوْجِ الْهَوَاءِ، وَذَلِكَ سِرٌّ لَا يَعْلَمُ حَقِيْقَتَهُ إِلَّا الْبَارِي سُبْحَانَهُ، إِلَى غَيْرِ ذَلِكَ.

ثُمَّ انْظُرْ كَيْفَ رَفَعَ الْأَنْفَ فِي وَسَطِ الْوَجْهِ، فَأَحْسَنَ شَكْلَهُ، وَفَتَحَ مَنْخِرَيْهِ، وَجَعَلَ فِيْهِمَا حَاسَّةَ الشَّمِّ؛ لِيَسْتَدِلَّ بِاسْتِنْشَاقِهِ عَلَى رَوَائِحِ مَطَاعِمِهِ وَمَشَارِبِهِ، وَلِيَتَنَعَّمَ بِالرَّوَائِحِ العَطِرَةِ وَيَتَجَنَّبَ الْخَبَائِثَ الْقَذِرَةَ، وَلِيَسْتَنْشِقَ أَيْضًا رُوْحَ الْحَيَاةِ؛ غِذَاءً لِقَلْبِهِ وَتَرْوِيْحاً لِحَرَارَةِ بَاطِنِهِ.

ثُمَّ خَلَقَ الْحَنْجَرَةَ وَهَيَّأَهَا لِخُرُوْجِ الْأَصْوَاتِ، وَدَوَّرَ اللِّسَانَ فِي

the essence of life as nourishment (i.e., oxygen) for his heart.[xli] (He made the two nostrils to allow) for the nourishment of his heart and ventilation of heat from the insides.

Then He created the throat, and made it facilitate the enunciation of voices. And (it also serves as an attachment) to rotate the tongue in movements and intonations so that the voice exits in (differently shaped) passages, such that the letters that come from it are also varied in dispersing the paths of speech. He made the throat with different shapes of narrowness and vastness, coarseness and smoothness, firmness and flaccidity, and lengthiness and shortness, such that men's voices differ because of these (shapes). So (observe that) two voices are never similar; just as He created differences between two faces (and they are not similar), two voices do not resemble each other.[xlii] Rather, the differences between two voices are apparent so that the listener can distinguish certain people from each other by way of nothing more than (listening to their) voice.[xliii] Likewise, differences between two faces are also apparent. That is a secret for recognition, for indeed when God the Exalted created Ādam and Eve, He made differences between their two faces.[xliv] Then He created from them a creation (their progeny) and made them different from the creation of their father and mother, and then each subsequent generation similarly followed (with different faces and voices). This is a secret for (mutual) recognition.[xlv]

Then look at the creation of the two hands and how they both guide (man) towards grasping intended (and useful) things and deflecting harmful things.[xlvi] (Look) at how He made the palm wide (through the arrangement of the fingers). He apportioned the five fingers and apportioned the fingers with finger bones (*anāmil*).[xlvii,xlviii] He made four fingers on one side (of the hand) and the thumb on the other so that it may hover and rotate over the other four fingers.[xlix,l] If the previous and later generations had assembled and attempted

الْحَرَكَاتِ وَالتَّقْطِيْعَاتِ، فَيَنْقَطِعُ الصَّوْتُ فِي مَجَارِي مُخْتَلِفَةٍ، تَخْتَلِفُ بِهَا الْحُرُوْفُ لِتَتَسِعَ طُرُقَ النُّطْقِ، وَجَعَلَ الْحَنْجَرَةَ مُخْتَلِفَةَ الْأَشْكَالِ فِي الضِّيْقِ وَالسَّعَةِ، وَالْخُشُوْنَةِ وَالْمَلَاسَةِ، وَصَلَابَةِ الْجَوْهَرِ وَرَخَاوَتِهِ، وَالطُّوْلِ وَالْقِصَرِ، حَتَّى اخْتَلَفَتْ بِسَبَبِ ذَلِكَ الْأَصْوَاتُ، فَلَمْ يَتَشَابَهْ صَوْتَانِ، كَمَا خَلَقَ بَيْنَ كُلِّ صُوْرَتَيْنِ اخْتِلَافًا فَلَمْ تَشْتَبِهْ صُوْرَتَانِ، بَلْ يَظْهَرُ بَيْنَ كُلِّ صُوْرَتَيْنِ فُرْقَانٌ، حَتَّى يُمَيِّزَ السَّامِعُ بَعْضَ النَّاسِ عَنْ بَعْضٍ بِمُجَرَّدِ الصَّوْتِ، وَكَذَلِكَ يَظْهَرُ بَيْنَ كُلِّ شَخْصٍ فُرْقَانٌ، وَذَلِكَ لِسِرِّ التَّعَارُفِ؛ فَإِنَّ اللهَ تَعَالَى لَمَّا خَلَقَ آدَمَ وَحَوَّاءَ خَالَفَ بَيْنَ صُوْرَتَيْهِمَا، فَخَلَقَ مِنْهُمَا خَلْقًا جَعَلَهُ مُخَالِفًا لِخَلْقِ أَبِيهِ وَأُمِّهِ، ثُمَّ تَوَالَى الْخَلْقُ كَذَلِكَ لِسِرِّ التَّعَارُفِ.

ثُمَّ انْظُرْ لِخَلْقِ الْيَدَيْنِ تَهْدِيَانِ إِلَى جَلْبِ الْمَقَاصِدِ وَدَفْعِ الْمَضَارِّ، وَكَيْفَ عَرَّضَ الْكَفَّ وَقَسَّمَ الْأَصَابِعَ الْخَمْسَ، وَقَسَّمَ الْأَصَابِعَ بِأَنَامِلَ، وَجَعَلَ الْأَرْبَعَةَ فِي جَانِبٍ وَالْإِبْهَامَ فِي جَانِبٍ آخَرَ، فَيَدُوْرُ الْإِبْهَامُ عَلَى الْجَمِيْعِ، فَلَوِ اجْتَمَعَ الْأَوَّلُوْنَ وَالْآخِرُوْنَ عَلَى أَنْ يَسْتَطِيْعُوا بِدِقَّةِ الْفِكْرِ وَجْهًا آخَرَ عَنْ وَضْعِ الْأَصَابِعِ سِوَى مَا وُضِعَتْ عَلَيْهِ مِنْ بُعْدِ الْإِبْهَامِ عَنِ الْأَرْبَعَةِ، وَتَفَاوُتِ الْأَرْبَعَةِ فِي الطُّولِ، وَتَرْتِيْبِهَا فِي صَفٍّ وَاحِدٍ

with meticulous planning to place (the fingers in an arrangement) other than how the fingers are placed, they would not (conceive of anything other) than an arrangement with distance between the thumb and the four fingers, variation in length of the four fingers, and their alignment in a single row.[li]

By virtue of this arrangement, the (tasks of) holding and giving become possible. If man spreads his fingers out, they become like a plate upon which he can place whatever he wishes. If he clenches them together (in a fist), they become a tool he may (use to) hit. If he brings them together, curved and not completely back (i.e., not touching the palm), the hand becomes like a ladle for him (to accomplish other tasks).[lii] If he stretches them out and keeps them joined together, it becomes like a scoop for him (to accomplish yet more tasks).[liii]

Then He created nails on the top (of fingers) as a beautification for the fingertips (*anāmil*) and as a support for them from behind. This is so that the fingers are not weakened by them, but rather become capable of picking up (even) minute things that are not capable of being reached if it were not for these fingernails.[liv] With their use, he can scratch his body when he has such a need. Ponder upon these small and minute things in his body: if he lacked them and an itch appeared, he would be the weakest of creation, incapable of fending off things that bothered him.[lv] So He brought about only that which would benefit him in that regard (to satisfy an itch), and there is nothing else that can replace fingernails in their ability to scratch his body, for indeed they were created for this and other purposes as well. So (look how) they are not hard like the hardness of bones, yet not flaccid like the flaccidity of skin.[lvi] It (the fingernail) lengthens and withers, and is trimmed and cut for certain purposes. He also made in man an ability to use his nails to guide his hand towards the source of itching – whether he is asleep or awake – and

لَمْ يَقْدِرُوا عَلَى ذَلِكَ، وَبِهَذَا الْوَضْعِ صَلَحَ بِهَا الْقَبْضُ وَالْإِعْطَاءُ، فَإِنْ بَسَطَهَا كَانَتْ طَبَقاً يَضَعُ عَلَيْهِ مَا يُرِيْدُ، وَإِنْ جَمَعَهَا كَانَتْ آلَةً يَضْرِبُ بِهَا، وَإِنْ ضَمَّهَا ضَمّاً غَيْرَ تَامٍّ كَانَتْ مِغْرَفَةً لَهُ، وَإِنْ بَسَطَهَا وَضَمَّ أَصَابِعَهُ كَانَتْ مِجْرَفَةً.

ثُمَّ خَلَقَ الْأَظَافِرَ عَلَى رُؤُوْسِهَا زِيْنَةً لِلْأَنَامِلِ، وَعِمَادًا لَهَا مِنْ وَرَائِهَا؛ حَتَّى لَا تَضْعُفَ، وَيَلْتَقِطَ بِهَا الْأَشْيَاءَ الدَّقِيْقَةَ الَّتِي لَا تَتَنَاوَلُهَا الْأَنَامِلُ لَوْلَاهَا، وَلِيَحُكَّ بِهَا جِسْمَهُ عِنْدَ الْحَاجَةِ إِلَى ذَلِكَ.

فَانْظُرْ أَقَلَّ الْأَشْيَاءِ فِي جِسْمِهِ لَوْ عَدِمَهَا وَظَهَرَتْ بِهِ حَكَّةٌ لَكَانَ أَضْعَفَ الْخَلْقِ وَأَعْجَزَهُمْ عَنْ دَفْعِ مَا يُؤْلِمُهُ، وَجَلْبِ مَا يَنْتَفِعُ بِهِ فِي ذَلِكَ، وَلَمْ يَقُمْ لَهُ غَيْرُ الظُّفْرِ مَقَامَهُ فِي حَكِّ جَسَدِهِ؛ لِأَنَّهُ مَخْلُوْقٌ لِذَلِكَ وَلِغَيْرِهِ، فَهُوَ لَا صَلْبٌ كَصَلَابَةِ الْعِظَامِ، وَلَا رَخْوٌ كَرَخَاوَةِ الْجِلْدِ، يَطُوْلُ وَيَخْلَقُ، وَيُقَصُّ وَيُقَصَّرُ لِمِثْلِ ذَلِكَ،

ثُمَّ جَعَلَهُ يَهْتَدِي بِهِ إِلَى الْحَكِّ فِي حَالَةِ نَوْمِهِ وَيَقَظَتِهِ، وَيَقْصِدُ الْمَوَاضِعَ إِلَى جِهَتِهَا مِنْ جَسَدِهِ، وَلَوِ احْتَاجَ إِلَى غَيْرِهِ وَاسْتَعَانَ بِهِ فِي حَكِّهَا لَمْ يَعْثُرِ الْغَيْرُ عَلَى مَوَاضِعِ الْحَاجَةِ إِلَّا بَعْدَ طُوْلٍ وَتَعَبٍ.

ثُمَّ أُنْظُرْ كَيْفَ مَدَّ مِنْهُ الْفَخِذَيْنِ وَالسَّاقَيْنِ وَبَسَطَ الْقَدَمَيْنِ؛

it reaches these places from any direction in his body.[lvii,lviii] If he had to resort to something else (i.e., some other organ or instrument) and sought aid with this to (accomplish) his scratching, this other thing would never come upon the places of need (i.e., the places that itch) except after much time and fatigue.[lix]

Then look at how He stretched out his thighs and shins for him and spread out his two feet to have the power and ability to walk. He adorned the feet with toes and made these as (both) a beautification and strength for walking.[lx] He further beautified and strengthened the toes with nails (as He did for the fingers).

Then look at how God created all of this from a despised drop of sperm; then He created the bones of the body from it, making them strong and solid forms in order to support the body and serve as pillars for it. He (the Most Blessed and Exalted) apportioned them with different measures, appropriate shapes, and forms. From amongst them are (bones) small and long; circular, hollow, solid, broad, and delicate.[lxi] Then He placed a delicate and preserved marrow in the hollows of these bones for their health and strengthening.[lxii]

Because man is in need of all of his body – and of some of his limbs more frequently due to his needs for them – God (may He be Glorified) did not make his bones as one bone, but rather as many. There are joints between bones such that movement becomes easy with them. He ordered the shape of each of them to be in accordance with the type of movement sought with each limb. Then He brought together the joint spaces and tied together some (of them) with others by (using) pegs (i.e., ligaments). He strengthened them from one end of the bone and attached the other side like a bandage. Then He created a protruding appendage on one end of the bone, and a recessed cavity in the other end of an (adjacent) bone that matches the shape of the protruding appendage. This is so that each (bone) fits suitably with the other (bone).[lxiii] So when man wishes to move one

لِيَتَمَكَّنَ بِذَلِكَ مِنَ السَّعْيِ، وَزَيَّنَ الْقَدَمَيْنِ بِالْأَصَابِعِ، وَجَعَلَهَا زِيْنَةً وَقُوَّةً عَلَى السَّعْيِ، وَزَيَّنَ الْأَصَابِعَ أَيْضاً بِالْأَظَافِرِ وَقَوَّاهَا بِهَا.

ثُمَّ أُنْظُرْ كَيْفَ خَلَقَ اللهُ هَذَا كُلَّهُ مِنْ نُطْفَةٍ مَهِيْنَةٍ، ثُمَّ خَلَقَ مِنْهَا عِظَامَ جَسَدِهِ، فَجَعَلَهَا أَجْسَامًا قَوِيَّةً صَلْبَةً؛ لِتَكُوْنَ قِواماً لِلْبَدَنِ وَعِمَادًا لَهُ، وَقَدَّرَهَا تَبَارَكَ وَتَعَالَى بِمَقَادِيْرَ مُخْتَلِفَةٍ وَأَشْكَالٍ مُتَنَاسِبَةٍ، فَمِنْهَا صَغِيْرٌ وَطَوِيْلٌ، وَمُسْتَدِيْرٌ وَمُجَوَّفٌ وَمُصْمَتٌ، وَعَرِيْضٌ وَدَقِيْقٌ. ثُمَّ أَوْدَعَ فِي أَنابِيْبِ هَذِهِ الْعِظَامِ الْمُخَّ الرَّقِيْقَ مُصَانًا لِمَصْلَحَتِهَا وَتَقْوِيَتِهَا.

وَلَمَّا كَانَ الْإِنْسَانُ مُحْتَاجاً إِلَى جُمْلَةِ جَسَدِهِ وَبَعْضِ أَعْضَائِهِ لِتَرَدُّدِهِ فِي حَاجَاتِهِ لَمْ يَجْعَلِ اللهُ سُبْحَانَهُ عِظَامَهُ عَظْماً وَاحِداً، بَلْ عِظَاماً كَثِيْرَةً وَبَيْنَهَا مَفَاصِلُ؛ حَتَّى تَتَيَسَّرَ بِهَا الْحَرَكَةُ، فَقَدَّرَ شَكْلَ كُلِّ وَاحِدَةٍ مِنْهَا عَلَى قَدْرٍ، وَفْقَ الْحَرَكَةِ الْمَطْلُوْبَةِ بِهَا، ثُمَّ وَصَلَ مَفَاصِلَهَا وَرَبَطَ بَعْضَهَا بِبَعْضٍ بِأَوْتَارٍ أَثْبَتَهَا بِأَحَدِ طَرَفَيِ الْعَظْمِ، وَأَلْصَقَ الطَّرَفَ الْآخَرَ كَالرِّبَاطِ، ثُمَّ خَلَقَ فِي أَحَدِ طَرَفَيِ الْعَظْمِ زَوَائِدَ خَارِجَةً مِنْهَا، وَمِنَ الْآخَرِ نُقَراً غَائِصَةً فِيْهَا تُوَافِقُ لِأَشْكَالِ الزَّوَائِدِ لِتَدْخُلَ فِيْهَا وَتَنْطَبِقَ، فَصَارَ الْإِنْسَانُ إِذَا أَرَادَ أَنْ يُحَرِّكَ شَيْئاً مِنْ جَسَدِهِ دُونَ غَيْرِهِ لَمْ يَمْتَنِعْ عَلَيْهِ، فَلَوْلَا حِكْمَةُ خَلْقِ الْمَفَاصِلِ لَتَعَذَّرَ عَلَيْهِ ذَلِكَ.

part of his body without moving another, he is not (anatomically) prevented from doing so.[lxiv] Had it not been for the Divine Wisdom of creating (these) joint spaces (and arrangements), you would have been incapable of this (movement).

Then look at how He made the creation of the head, comprised of 55 bones of different forms and shapes.[lxv] He joined some of them to others such that the skull is perfectly balanced (atop the spine), as is apparent.[lxvi] Six bones from these (55 bones) are exclusively for the skull, 24 (bones) are for the upper jaw, and two (bones) are for the lower jaw.[lxvii] The rest are from the teeth; some are broad and flat and suitable for grinding (food), and some are sharp and suitable for cutting (food).[lxviii]

Then He made the neck as a seat for the head. He assembled it from seven hollow circular vertebrae, having (some portions of) excessiveness and defectiveness, so that some are suitable for (articulating more appropriately) with others – to mention (the amount of) Divine Wisdom on this matter would be quite lengthy.[lxix,lxx] Then He constructed the neck upon the back (i.e., the spine), from the bottom of the neck until the end of the sacral bone, (consisting of) 24 vertebrae and the sacral bone (joined) with three other (bones).[lxxi] He joined the coccyx bone, which itself is composed of three other bones, to the bottom (of these vertebrae).[lxxii,lxxiii] Then He joined the spine with the pelvic bones, the shoulder bones (i.e., the clavicle and scapula), the bones of the two arms, the bones of the sacrum, and the bones of the thighs (femurs), shins, and bones of the feet. He made the total number of bones in the human body as 248, not including the small bones that complete the intervening joint spaces.[lxxiv]

Look at how the Inventor (may He be Glorified and Exalted) created all of that from a delicate and insignificant fluid. The sole intent in mentioning the number of bones is to demonstrate the

ثُمَّ انْظُرْ كَيْفَ جَعَلَ خَلْقَ الرَّأْسِ مُرَكَّباً مِنْ خَمْسَةٍ وَخَمْسِيْنَ عَظْماً مُخْتَلِفَةَ الْأَشْكَالِ وَالصُّوَرِ، وَأَلَّفَ بَعْضَهَا إِلَى بَعْضٍ بِحَيْثُ اسْتَوَتْ كُرَةُ الرَّأْسِ كَمَا تَرَى، فَمِنْهَا سِتَّةٌ تَخْتَصُّ بِالقِحْفِ، وَأَرْبَعَةٌ وَعِشْرُوْنَ لِلَّحْيِ الْأَعْلَى، وَاثْنَانِ لِلَّحْيِ الْأَسْفَلِ، وَالْبَقِيَّةُ مِنَ الْأَسْنَانِ بَعْضُهَا عَرِيْضٌ يَصْلُحُ لِلطَّحْنِ، وَبَعْضُهَا حَادٌّ يَصْلُحُ لِلْقَطْعِ.

ثُمَّ جَعَلَ الرَّقَبَةَ مَرْكَزَ الرَّأْسِ، فَرَكَّبَهَا مِنْ سَبْعِ خَرَزَاتٍ مُجَوَّفَاتٍ مُسْتَدِيْرَاتٍ وَزِيَادَاتٍ وَنُقْصَانٍ لِيَنْطَبِقَ بَعْضُهَا عَلَى بَعْضٍ، وَيَطُوْلُ ذِكْرُ الْحِكْمَةِ فِيْهَا. ثُمَّ رَكَّبَ الرَّقَبَةَ عَلَى الظَّهْرِ مِنْ أَسْفَلِ الرَّقَبَةِ إِلَى مُنْتَهَى عَظْمِ الْعَجُزِ مِنْ أَرْبَعَةٍ وَعِشْرِيْنَ خَرَزَةً، وَعَظْمُ الْعَجُزِ ثَلَاثَةٌ أُخْرَى مُخْتَلِفَةٌ، وَوَصَلَ بِهِ مِنْ أَسْفَلِهِ عَظْمَ الْعُصْعُصِ، وَهُوَ مُؤَلَّفٌ مِنْ ثَلَاثَةٍ أُخْرَى، ثُمَّ وَصَلَ عِظَامَ الظَّهْرِ بِعِظَامِ الصَّدْرِ وَعِظَامِ الْكَتِفِ وَعِظَامِ الْيَدَيْنِ وَعِظَامِ الْعَانَةِ وَعِظَامِ الْعَجُزِ وَعِظَامِ الْفَخِذَيْنِ وَالسَّاقَيْنِ وَأَصَابِعِ الرِّجْلَيْنِ، فَجَعَلَ جُمْلَةَ عَدَدِ الْعِظَامِ فِي بَدَنِ الْإِنْسَانِ مِائَتَيْ عَظْمٍ وَثَمَانِيَةً وَأَرْبَعِيْنَ عَظْماً، سِوَى الْعِظَامِ الصَّغِيْرَةِ الَّتِي حَشَا بِهَا خَلَلَ الْمَفَاصِلِ. فَانْظُرْ كَيْفَ خَلَقَ الْبَارِي سُبْحَانَهُ وَتَعَالَى ذَلِكَ كُلَّهُ مِنْ نُطْفَةٍ رَقِيْقَةٍ سَخِيْفَةٍ، وَالْمَقْصُوْدُ مِنْ ذِكْرِ أَعْدَادِهَا تَعْظِيْمُ مُدَبِّرِهَا

grandness of His Design and Creation, how He created them, how He differentiated them according to their shapes, and how He distinguished them with a special measurement.[lxxv] If only one extra bone was added (in an unsuitable place), it would cause an ailment (in the human being), and he would need to remove it. If he were missing or having a defect in any (bone), he would have needed to compensate for it. He (may He be Glorified and Exalted) made a lesson in creation for people of perspicacity, and (He provided) clearly evident signs to demonstrate His Greatness and Loftiness concerning His Measurement and Shaping.[lxxvi]

Then look at how He (Glorified is He) created muscles as instruments for moving the bones. He created 529 muscles in the human body.[lxxvii] Muscles are constructed from flesh, then nerves and membranes are joined and covered (over them). They are of different capacities and shapes in accordance with their placement and need. So, 24 (muscles) are for the movement of the eyes and eyelids alone, such that if even one were missing or defective, the movement of the eye would become unbalanced and defective.[lxxviii] Likewise, there are muscles appropriately numbered for each limb and organ, being exclusively for them and measured in perfect accordance and agreement (with their functions).[lxxix]

As for the matter concerning the nerves, veins, arteries, arterioles, their proxies, and their volumes, I am in wonderment at them, and their explanation is lengthy. As for the amazements contained in the (subtle and hidden) meanings and qualities (of these tissues), these are not understood by the senses, and are far greater.[lxxx,lxxxi]

Then look at all that has been ennobled and specified exclusively in his creation, such that he was created being able to rise up by standing and equally sit down (without difficulty).[lxxxii] So he takes on matters and affairs with his hands and limbs and is capable of (tasks such as) handling and working.[lxxxiii] And (marvel that) he was not

وَخَالِقِهَا، وَكَيْفَ خَلَقَهَا وَخَالَفَ بَيْنَ أَشْكَالِهَا، وَخَصَّهَا بِهَذَا الْقَدْرِ الْمَخْصُوْصِ؛ بِحَيْثُ لَوِ ازْدَادَ فِيْهَا عَظْمٌ وَاحِدٌ لَكَانَ وَبَالاً، وَاحْتَاجَ الْإِنْسَانُ إِلَى قَلْعِهِ، وَلَوْ نَقَصَ مِنْهَا وَاحِدٌ لَاحْتَاجَ إِلَى جَبْرِهِ، فَجَعَلَ سُبْحَانَهُ وَتَعَالَى فِي هَذَا الْخَلْقِ عِبْرَةً لِأُولِي الْأَبْصَارِ، وَآيَاتٍ بَيِّنَاتٍ عَلَى عَظَمَتِهِ وَجَلَالِهِ بِتَقْدِيْرِهَا وَتَصْوِيْرِهَا.

ثُمَّ انْظُرْ كَيْفَ خَلَقَ سُبْحَانَهُ آلَاتٍ لِتَحْرِيْكِ الْعِظَامِ وَهِيَ الْعَضَلَاتُ، فَخَلَقَ فِي بَدَنِ الْإِنْسَانِ خَمْسَمِائَةٍ وَتِسْعَةٍ وَعِشْرِيْنَ عَضَلَةً، وَالعَضَلَةُ مُرَكَّبَةٌ مِنْ لَحْمٍ وَعَصَبٍ وَرِبَاطٍ وَأَغْشِيَةٍ، وَهِيَ مُخْتَلِفَةُ الْمَقَادِيْرِ وَالْأَشْكَالِ بِحَسَبِ اخْتِلَافِ مَوَاضِعِهَا وَحَاجَتِهَا، فَأَرْبَعَةٌ وَعِشْرُوْنَ مِنْهَا لِحَرَكَةِ الْعَيْنِ وَأَجْفَانِهَا، بِحَيْثُ لَوْ نَقَصَتْ مِنْهَا وَاحِدَةٌ اخْتَلَّ أَمْرُ الْعَيْنِ، وَهَكَذَا لِكُلِّ عُضْوٍ عَضَلَاتٌ بِعَدَدٍ يَخُصُّهُ وَقَدْرٌ يُوَافِقُهُ.

وَأَمَّا أَمْرُ الْأَعْصَابِ وَالْعُرُوْقِ وَالْأَوْرِدَةِ وَالشَّرَايِيْنِ وَمَنَابِتِهَا وَسَعَتِهَا فَأَعْجَبُ مِنْ هَذَا، وَشَرْحُهُ يَطُوْلُ. ثُمَّ عَجَائِبُ مَا فِيْهِ مِنَ الْمَعَانِي وَالصِّفَاتِ الَّتِي لَا تُدْرَكُ بِالْحَوَاسِّ أَعْظَمُ.

ثُمَّ انْظُرْ إِلَى مَا شُرِّفَ بِهِ وَخُصِّصَ فِي خَلْقِهِ بِأَنَّهُ خُلِقَ يَنْتَصِبُ قَائِماً، وَيَسْتَوِي جَالِساً، وَيَسْتَقْبِلُ الْأُمُوْرَ بِيَدَيْهِ وَجَوَارِحِهِ، وَيُمْكِنُهُ

created in a prostrate position upon his face like some animals, for had this been (his state), he would be incapable of (accomplishing) these tasks.

Then look comprehensively at the outward and inner appearance of man. You will find (how) it is manufactured with such Divine Wisdom that it necessitates amazement.[lxxxiv] Indeed, He made his limbs completely capable of procuring nourishment, and (made) nourishment itself in a form that befits him. He (the Most Blessed and Exalted) created them with fixed proportions that are not enlarged. Rather, they remain (upon their size and shape) and do not increase (in size) after consumption of more food.[lxxxv] If the limbs remained continuously increasing because of (the consumption of) food, then indeed the bodies of the children of Ādam ﷺ would become too large and heavy to move, and would be incapable of (performing) fine and skilled work.[lxxxvi] Nothing reaches (man) from nourishment except that which is beneficial to him – his clothes and places of dwelling are likewise (only to benefit him). All of this is as a result of considerable Divine Wisdom and excellent Planning, (such that even) its limits are upon a Decreed boundary.[lxxxvii] (All of this is) a Divine Mercy and Kindness from God for His creation.

So, when you find that all of this is the creation of God the Exalted from a single drop of water, then what do you think of His creation in the kingdom of the heavens, the expanses of the Earth, and in the Sun, Moon, and planets?[lxxxviii] What (do you think) of His Wisdom in their measurement, shapes, numbers, placements, and the togetherness of some and separation of others? (What do you think of) these different forms and the differences of (the universe's) easts and wests? So do not suspect that even the smallest speck in the heavens, Earth, and the rest of God's Knowledge is devoid of Divine Wisdom! Rather, (know) that (every creation) is inclusive of all (possible) wonderment and wisdoms, and none can truly

الْعِلَاجُ وَالْعَمَلُ، وَلَمْ يُخْلَقْ مَكْبُوْباً عَلَى وَجْهِهِ كَعِدَّةٍ مِنَ الْحَيَوَانَاتِ؛ إِذْ لَوْ كَانَ كَذَلِكَ لَمَا اسْتَطَاعَ هَذِهِ الْأَعْمَالَ.

ثُمَّ انْظُرْ مِنْ حَيْثُ الْجُمْلَةُ إِلَى ظَاهِرِ الْإِنْسَانِ وَبَاطِنِهِ فَتَجِدُهُ مَصْنُوْعاً صَنْعَةً بِحِكْمَةٍ تَقْضِيْ مِنْهَا الْعَجَبَ، وَقَدْ جَعَلَ سُبْحَانَهُ أَعْضَاءَهُ تَامَّةً بِالْغِذَاءِ، وَالْغِذَاءُ مُتَوَالٍ عَلَيْهَا، لَكِنَّهُ تَبَارَكَ وَتَعَالَى قَدَّرَهَا بِمَقَادِيرَ لَا يَتَعَدَّاهَا، بَلْ يَقِفُ عِنْدَهَا وَلَا يَزِيْدُ عَلَيْهَا، فَإِنَّهَا لَوْ تَزَايَدَتْ بِتَوَالِي الْغِذَاءِ عَلَيْهَا لَعَظُمَتْ أَبْدَانُ بَنِيْ آدَمَ وَثَقُلَتْ عَنِ الْحَرَكَةِ، وَعُطِّلَتْ عَنِ الصِّنَاعَاتِ اللَّطِيْفَةِ، وَلَا تَنَاوَلَتْ مِنَ الْغِذَاءِ مَا يُنَاسِبُهَا، وَمِنَ اللِّبَاسِ كَذَلِكَ، وَمِنَ الْمَسَاكِنِ مِثْلَ ذَلِكَ، وَكَانَ مِنْ بَلِيْغِ الْحِكْمَةِ وَحُسْنِ التَّدْبِيْرِ وُقُوْفُهَا عَلَى هَذَا الْحَدِّ الْمُقَدَّرِ رَحْمَةً مِنَ اللهِ وَرِفْقاً بِخَلْقِهِ.

فَإِذَا وَجَدْتَ هَذَا كُلَّهُ صَنْعَةَ اللهِ تَعَالَى مِنْ قَطْرَةِ مَاءٍ، فَمَا ظَنُّكَ بِصَنْعَتِهِ فِي مَلَكُوْتِ السَّمَاوَاتِ وَالْأَرْضِ وَشَمْسِهَا وَقَمَرِهَا وَكَوَاكِبِهَا؟ وَمَا حِكْمَتُهُ فِي أَقْدَارِهَا وَأَشْكَالِهَا وَأَعْدَادِهَا وَأَوْضَاعِهَا، وَاجْتِمَاعِ بَعْضِهَا وَافْتِرَاقِ بَعْضِهَا، وَاخْتِلَافِ صُوَرِهَا، وَتَفَاوُتِ مَشَارِقِهَا وَمَغَارِبِهَا؟ فَلَا تَظُنَّ أَنَّ ذَرَّةً فِي السَّمَاوَاتِ وَالْأَرْضِ وَسَائِرِ عِلْمِ اللهِ

comprehend all of this save God, Glorified and Exalted is He.[lxxxix] Have you not heard His saying (may He be Glorified and Exalted): "*Are you a more difficult creation or the heavens – He created it!*"[xc] until the end of (the verse) that He, the Exalted, has indicated.

Consider that if man and jinn were to gather together (and attempt) to create from sperm (something) that can hear, see, and possess life, they would never be capable of that. Then look at how He (may He be Glorified) created (the faculties of hearing and seeing) in the womb and shaped them, making excellent their fashioning.[xci] He measured them and made excellent their measuring, He formed them and made excellent their forming, and He apportioned their similar parts to different parts. Then He consolidated the bones, delaying them (in their appearance in the fetus). He beautified the shapes of the limbs, arranging their veins and nerves.[xcii] He organized his exterior and interior, and made a pathway for his nourishment between them so that it would be a means for his survival for the duration of his entire lifetime.

Then (look) how He arranged the internal organs, including the heart, liver, stomach, spleen, lungs, uterus, bladder, and intestines. Each organ is upon a specified form, (arranged) with a specific measurement for the performance of a specific task. He made the stomach as durable, strong sinew[xciii] to (facilitate its function of) digesting nourishment with its requirement (of such a sturdy construction) for this (digestion). Thus, it is possible (for the stomach) to cut and grind (food).[xciv,xcv] He made the initial chewing (of food) by the teeth as an aid for the stomach, in order to achieve efficiency in (both) grinding and digesting.[xcvi] He made the liver for the purpose of diverting (digested) nourishment to the blood, such that blood may draw from it (and transport it) to every single organ (the specific) nourishment that is most suitable for it.[xcvii] The nourishment of the bones is different from the nourishment of the

يَنْفَكُّ عَنْ حِكَمٍ، بَلْ ذَلِكَ مُشْتَمِلٌ عَلَى عَجَائِبَ وَحِكَمٍ لَا يُحِيْطُ بِجَمِيْعِهَا إِلَّا اللهُ سُبْحَانَهُ وَتَعَالَى، أَلَمْ تَسْمَعْ قَوْلَهُ سُبْحَانَهُ وَتَعَالَى: {أَأَنتُمْ أَشَدُّ خَلْقاً أَمِ السَّمَاءُ بَنَاهَا} [النازعات: ٢٧] إِلَى آخِرِ مَا نَبَّهَ بِهِ تَعَالَى.

وَتَأَمَّلْ لَوِ اجْتَمَعَ الْإِنْسُ وَالْجِنُّ عَلَى أَنْ يَخْلُقُوا لِلنُّطْفَةِ سَمْعاً وَبَصَراً وَحَيَاةً لَمْ يَقْدِرُوا عَلَى ذَلِكَ، فَانْظُرْ كَيْفَ خَلَقَهَا سُبْحَانَهُ فِي الْأَرْحَامِ، وَشَكَّلَهَا فَأَحْسَنَ تَشْكِيْلَهَا، وَقَدَّرَهَا فَأَحْسَنَ تَقْدِيْرَهَا، وَصَوَّرَهَا فَأَحْسَنَ تَصْوِيرَهَا، وَقَسَّمَ أَجْزَاءَهَا الْمُتَشَابِهَةَ إِلَى أَجْزَاءٍ مُخْتَلِفَةٍ، فَأَحْكَمَ الْعِظَامَ فِي أَرْجَائِهَا، وَحَسَّنَ أَشْكَالَ أَعْضَائِهَا، وَرَتَّبَ عُرُوْقَهَا وَأَعْصَابَهَا، وَدَبَّرَ ظَاهِرَهَا وَبَاطِنَهَا، وَجَعَلَ فِيْهَا مَجْرَى لِغِذَائِهَا؛ لِيَكُوْنَ ذَلِكَ سَبَبًا لِبَقَائِهَا مُدَّةَ حَيَاتِهَا.

ثُمَّ كَيْفَ رَتَّبَ الْأَعْضَاءَ الْبَاطِنَةَ مِنَ الْقَلْبِ وَالْكَبِدِ وَالْمَعِدَةِ وَالطِّحَالِ وَالرِّئَةِ وَالرَّحِمِ وَالْمَثَانَةِ وَالْأَمْعَاءِ، كُلُّ عُضْوٍ بِشَكْلٍ مَخْصُوْصٍ وَمِقْدَارٍ مَخْصُوْصٍ لِعَمَلٍ مَخْصُوْصٍ، فَجَعَلَ الْمَعِدَةَ لِنُضْجِ الْغِذَاءِ عَصَباً مَتِيْناً شَدِيْداً لِحَاجَتِهَا إِلَى ذَلِكَ، وَبِذَلِكَ يُمْكِنُ تَقْطِيْعُهُ وَطَحْنُهُ، وَجَعَلَ طَحْنَ الْأَضْرَاسِ أَوَّلاً مُعِيْناً لِلْمَعِدَةِ عَلَى جَوْدَةِ طَحْنِهِ وَهَضْمِهِ، وَجَعَلَ الْكَبِدَ لِإِحَالَةِ الْغِذَاءِ إِلَى الدَّمِ، فَيَجْذِبُ مِنْهُ إِلَى كُلِّ عُضْوٍ مِنَ

muscles and flesh, and that of the vasculature is different from that of the nerves. And the nourishment of hair is different from other organs.[xcviii] He made the spleen, gallbladder, and kidneys in order to serve the liver. The spleen attracts the black (bile) and the gallbladder attracts the yellow (bile).[xcix] The kidney draws wetness (liquid waste) from the human, and the bladder exists to accept the liquid waste that is transported away from the kidney. This waste exits (the body) by passing through the urethra. The vasculature then transports filtered blood (from the kidneys) to the rest of the body. He made the composition (of the bladder or kidneys) to be stronger than the composition of flesh, such that they may receive and contain urine. Thus, they were made to resemble containers and vessels.[c]

Then look at how He prepared him in the womb and enveloped him with kindness – this discussion is lengthy, and its knowledge is not completely known except by its Creator. The describer (i.e., the human being) is incapable of elucidating (the knowledge that) reaches him with only a cursory observation of that (matter). But (one thing he may describe is that) his term in it is such that he has no need to call out (while in the womb).[ci]

Therefore, the newborn does not need to be taught – neither through exhortation nor through warning – how to accomplish (his needs). Rather, it is the newborn's instinct that when there is the need to cry out for his feeding (he knows how to do so).[cii] Had it not been for that (crying), mothers would eschew their newborn children due to the severe tiredness and hardship of (his) upbringing. It is only when his body becomes heavier, his outer and inner organs develop, and he becomes capable of digesting food that He brings forth (his) teeth, that is, only when he is in need of them; neither before nor after this.[ciii,civ]

Then look how God created the faculties of perception and intelligence in gradual degrees until his completion and maturity.

الْغِذَاءِ مَا يُنَاسِبُهُ، فَغِذَاءُ الْعَظْمِ خِلَافُ غِذَاءِ اللَّحْمِ، وَغِذَاءُ الْعُرُوْقِ خِلَافُ غِذَاءِ الْأَعْصَابِ، وَغِذَاءُ الشَّعْرِ خِلَافُ غِذَاءِ غَيْرِهِ، وَجَعَلَ الطِّحَالَ وَالْمَرَارَةَ وَالْكُلْيَةَ لِخِدْمَةِ الْكَبِدِ، فَالطِّحَالُ لِجَذْبِ السَّوْدَاءِ، وَالْمَرَارَةُ لِجَذْبِ الصَّفْرَاءِ، وَالْكُلْيَةُ لِجَذْبِ الْمَاءِ عَنْهُ، وَالْمَثَانَةُ لِقَبُوْلِ الْمَاءِ عَنْ الْكُلْيَةِ، ثُمَّ يُخْرِجُهُ فِي مَجْرَى الْإِحْلِيْلِ، وَالْعُرُوْقُ لِاتِّصَالِ الدَّمِ مِنْهَا إِلَى سَائِرِ أَطْرَافِ الْبَدَنِ، وَجَعَلَ جَوْهَرَهَا أَتْقَنَ مِنْ جَوْهَرِ اللَّحْمِ؛ لِتَصُوْنَ الدَّمَ وَتَحْصُرَهُ، فَهِيَ بِمَنْزِلَةِ الظُّرُوْفِ وَالْأَوْعِيَةِ.

ثُمَّ انْظُرْ كَيْفَ دَبَّرَهُ فِي الرَّحِمِ، وَلَطَفَ بِهِ أَلْطَافاً يَطُوْلُ شَرْحُهَا، وَلَا يَسْتَكْمِلُ الْعِلْمَ بِجُمْلَتِهَا إِلَّا خَالِقُهَا، وَيَعْجِزُ الْوَاصِفُ عَنْ وَصْفِ مَا وَصَلَ إِلَيْهِ نَظَرُهُ مِنْ ذَلِكَ، فَمِنْ ذَلِكَ جَعْلُهُ فِيْهِ لَا يَحْتَاجُ إِلَى اسْتِدْعَاءٍ، وَلَا يَحْتَاجُ الْمَوْلُوْدُ إِلَى مَا يُبَيِّنُ لَهُ ذَلِكَ، لَا بِوَعْظٍ وَلَا تَنْبِيْهٍ، بَلْ ذَلِكَ فِي الطِّبَاعِ إِلَى وَقْتِ حَاجَةِ الْمَوْلُوْدِ إِلَى الْإِغَاثَةِ فِي غِذَائِهِ، وَلَوْلَا ذَلِكَ لَنَفَرَتِ الْأُمَّهَاتُ عَنْهُ مِنْ شِدَّةِ التَّعَبِ وَكُلْفَةِ التَّرْبِيَةِ، حَتَّى إِذَا اشْتَدَّ جِسْمُهُ وَقَوِيَتْ أَعْضَاؤُهُ الظَّاهِرَةُ وَالْبَاطِنَةُ لِهَضْمِ الْغِذَاءِ، فَحِيْنَئِذٍ أَنْبَتَ لَه الْأَسْنَانَ عِنْدَ الْحَاجَةِ إِلَيْهَا لَا قَبْلَ ذَلِكَ وَلَا بَعْدَهُ. ثُمَّ انْظُرْ كَيْفَ خَلَقَ اللهُ فِيْهِ التَّمْيِيْزَ وَالْعَقْلَ عَلَى التَّدْرِيْجِ إِلَى حِيْنِ كَمَالِهِ

Look and ponder upon the essence of his being; he was born ignorant, without any intelligence or understanding. Indeed, if he were born as a child possessing intelligence and understanding, he would be overwhelmed by the (surrounding) world from the time of his exit (from the womb); in such a case, he would remain bewildered, his intelligence wandering (in contemplation) since he would see things he would not recognize, and experience that which he had not yet seen or grown accustomed to.[cv] He would find an imperfection upon seeing himself carried and lodged, wrapped with cloth, laid out and blanketed in the cradle, in spite of the fact that he is not devoid of needing all this due to the flimsiness of his body and his wetness upon being born. (Had the child possessed such faculties immediately) the softness, sweetness, and love that is naturally present in the hearts for young ones would not exist due to (the child's) many objections (that would arise) from his "intelligence" and choice (of certain things or actions) for himself. It is therefore evident that the incremental growth of man's intelligence and comprehension by degrees is most suitable for him. Is it not evident that He set up everything from his creation with Divine Wisdom and intended a proper course (for him)? He also informed him about the various errors in His minuteness and sublimity.

Then consider how, when he becomes strong (enough), He created a path and reason for procreation in him. He created facial hair in him to distinguish him from resembling young children and women, for it beautifies him and covers him from the wrinkles of his face when he becomes old. If she is a female, her face remains devoid of hair so that beauty and youthfulness remain in her, serving as an attraction for men so that procreation may continue.

Ponder now on what we have mentioned and what He (Glorified is He) arranged (in the human) in all these different conditions. Do you think something (as complicated) as this is possible or easy?[cvi]

وَبُلُوْغِهِ، وَانْظُرْ وَفَكِّرْ فِي سِرِّ كَوْنِهِ يُوْلَدُ جَاهِلاً غَيْرَ ذِيْ عَقْلٍ وَفَهْمٍ، فَإِنَّهُ لَوْ كَانَ وُلِدَ عَاقِلاً فِيْهَا لَأَنْكَرَ الْوُجُوْدَ عِنْدَ خُرُوْجِهِ إِلَيْهِ حَتَّى يَبْقَى حَيْرَانَ تَائِهَ الْعَقْلِ؛ إِذْ رَأَى مَا لَا يَعْرِفُ، وَوَرَدَ عَلَيْهِ مَا لَمْ يَرَهُ وَلَمْ يَعْهَدْ مِثْلَهُ، ثُمَّ كَانَ يَجِدُ غَضَاضَةً أَنْ يَرَى نَفْسَهُ مَحْمُوْلاً وَمَوْضُوْعاً مُعَصَّباً بِالْخِرَقِ وَمُسَجّىً فِي الْمَهْدِ، مَعَ كَوْنِهِ لَا يَسْتَغْنِي عَنْ هَذَا كُلِّهِ لِرِقَّةِ بَدَنِهِ وَرُطُوْبَتِهِ حِيْنَ يُوْلَدُ، ثُمَّ كَانَ لَا يُوْجَدُ لَهُ مِنَ الرِّقَّةِ وَالْحَلَاوَةِ وَالْمَحَبَّةِ فِي الْقُلُوبِ مَا يُوْجَدُ لِلصَّغِيرِ لِكَثْرَةِ اعْتِرَاضِهِ بِعَقْلِهِ وَاخْتِيَارِهِ لِنَفْسِهِ، فَتَبَيَّنَ أَنَّ ازْدِيَادَ الْعَقْلِ وَالْفَهْمِ فِيْهِ عَلَى التَّدْرِيجِ أَصْلَحُ بِهِ. أَفَلَا يُرَى كَيْفَ أَقَامَ كُلَّ شَيْءٍ فِيْهِ مِنَ الْخِلْقَةِ عَلَى غَايَةِ الْحِكْمَةِ وَطَرِيقِ الصَّوَابِ؟ وَأَعْلَمَهُ تَقَلُّبَ الْخَطَأِ فِي دَقِيْقِهِ وَجَلِيْلِهِ.

ثُمَّ انْظُرْ فِيْمَا إِذَا اشْتَدَّ خَلَقَ فِيْهِ طَرِيْقاً وَسَبَباً لِلتَّنَاسُلِ، وَخَلَقَ فِي وَجْهِهِ شَعْراً لِيُمَيِّزَهُ عَنْ شِبْهِ الصِّبْيَانِ وَالنِّسْوَانِ، وَيُجَمِّلَهُ وَيَسْتُرَ بِهِ غُضُوْنَ وَجْهِهِ عِنْدَ شَيْخُوْخَتِهِ، وَإِنْ كَانَتْ أُنْثَى أَبْقَى وَجْهَهَا نَقِيّاً مِنَ الشَّعْرِ لِتَبْقَى لَهَا بَهْجَةٌ وَنَضَارَةٌ تُحَرِّكُ الرِّجَالَ؛ لِمَا فِي ذَلِكَ مِنْ بَقَاءِ النَّسْلِ. فَكِّرِ الْآنَ فِيْمَا ذَكَرْنَاهُ وَدَبَّرَهُ سُبْحَانَهُ فِي هَذِهِ الْأَحْوَالِ الْمُخْتَلِفَةِ، هَلْ تَرَى مِثْلَ هَذَا يُمْكِنُ أَنْ يَكُوْنَ مُهْمَلاً؟ أَرَأَيْتَ لَوْ

Do you not see that if blood did not course through his body – as nourishment while he is in the womb – would he not wither, die, and dry up like vegetation when it is cut off from water?[cvii] Had (his mother's) labor not disturbed him once he matured (in the womb), would he not then be destroyed by remaining in the womb, along with his mother?[cviii] Had milk been disagreeable to him after his birth, would he not die from hunger and thirst, or be forced to consume that which is unsuitable and unwholesome for his body?[cix] Had He not created teeth for him at an appropriate time, would he not be incapable of chewing food and swallowing it, and would (he not) remain suckling such that his body would never become strong? Had He not caused facial hair to grow on him, he would remain resembling women and children, so you would not see (signs of) dignity, grandeur, and respect in him. Who is the one that provides (all this) for him – until he is sufficed for every and all of his needs in their appropriate times – except the One who raised him after he was not even a remembered thing, then favored him, and blessed him with each of these blessings?[cx]

Ponder upon the concupiscence of intimate relations that serve as a motive for his procreation; and (ponder upon) the organs that transport sperm to the womb; and the requisite intimate movements to cause the release of sperm – how all of this resulted only from (His) Wise Deliberation. Then ponder intensely upon the limbs of the body and (see) how they have been prepared. Each limb is (created) for the purpose that is intended to be accomplished from it. The eyes are suited for guiding oneself through the faculty of vision, the hands are suited for tasks such as handling, throwing, and pushing, while the feet are suited for walking. The stomach is suited for the digestion of food, and the liver is suited for extracting and separating (the individual nutrients).[cxi] The mouth is designed for speech and as an entrance for food. (Various) orifices are designed to remove waste and

لَمْ يُجْرِ لَهُ الدَّمَ غِذَاءً وَهُوَ فِي الرَّحِمِ؟ أَلَمْ يَكُنْ يَذْوِي وَيَهْلِكُ وَيَجِفُّ كَمَا يَجِفُّ النَّبَاتُ إِذَا انْقَطَعَ عَنْهُ الْمَاءُ؟ وَلَوْ لَمْ يُزْعِجْهُ الْمَخَاضُ عِنْدَ اسْتِكْمَالِهِ، أَلَمْ يَكُنْ يَهْلِكُ بِبَقَائِهِ فِي الرَّحِمِ هُوَ وَأُمُّهُ؟ وَلَوْ لَمْ يُوَافِقْهُ اللَّبَنُ عِنْدَ وِلَادَتِهِ، أَلَمْ يَكُنْ يَمُوتُ جُوعاً وَعَطَشاً، أَو يُغَذَّى بِمَا لَا يُوَافِقُ وَلَا يَصْلُحُ عَلَيْهِ بَدَنُهُ؟ وَلَوْ لَمْ يَخْلُقْ لَهُ الْأَسْنَانَ فِي وَقْتِهَا، أَلَمْ يَكُنْ يَمْتَنِعُ عَلَيْهِ مَضْغُ الطَّعَامِ وَازْدِرَادُهُ؟ وَيُقِيمُ عَلَى الرَّضَاعِ، وَلَا يَشْتَدُّ جِسْمُهُ؟ وَلَوْ لَمْ يَخْرُجْ لَهُ شَعْرُ الْوَجْهِ لَبَقِيَ فِي هَيْئَةِ النِّسَاءِ وَالصِّبْيَانِ، فَلَا تَرَى لَهُ هَيْبَةً وَلَا جَلَالاً وَلَا وَقَاراً، وَمَنْ ذَا الَّذِي يَرْصُدُهُ حَتَّى يُوَفِّيَهُ بِكُلِّ هَذِهِ الْمَآرِبِ فِي وَقْتِهَا إِلَّا الَّذِي أَنْشَأَهُ بَعْدَ أَنْ لَمْ يَكُنْ شَيْئاً مَذْكُوراً؟ وَتَفَضَّلَ عَلَيْهِ وَمَنَّ عَلَيْهِ بِكُلِّ هَذِهِ النِّعَمِ.

فَكِّرْ فِي شَهْوَةِ الْجِمَاعِ الدَّاعِيَةِ لِإِحْيَائِهِ، وَالْآلَةِ الْمُوصِلَةِ إِلَى الرَّحِمِ النُّطْفَةَ، وَالْحَرَكَةِ الْمُوجِبَةِ لِاسْتِخْرَاجِ النُّطْفَةِ، وَمَا فِي ذَلِكَ مِنَ التَّدْبِيرِ الْمُحْكَمِ، ثُمَّ فَكِّرْ فِي جُمْلَةِ أَعْضَاءِ الْبَدَنِ، وَتَهْيِئَةِ كُلِّ عُضْوٍ مِنْهَا لِلْأَرَبِ الَّذِي أُرِيدَ مِنْهَا، فَالعَيْنَانِ لِلِاهْتِدَاءِ بِالنَّظَرِ، وَالْيَدَانِ لِلْعِلَاجِ وَالْحَذْفِ وَالدَّفْعِ، وَالرِّجْلَانِ لِلسَّعْيِ، وَالْمَعِدَةُ لِهَضْمِ الطَّعَامِ، وَالْكَبِدُ لِلتَّخْلِيصِ وَالتَّمْيِيزِ، وَالْفَمُ لِلْكَلَامِ وَدُخُولِ الْغِذَاءِ، وَالمَنَافِذُ لِدَفْعِ

excess.[cxii] Likewise, when you meditate about the rest of the wonders that are in the human being, you will conclude that indeed all (of these mechanisms) were placed with a wise aim and intent.

Ponder upon the transport of food to the stomach where it ripens, and how it sends these purified (and digested) products to the liver through fine veins constructed like a sieve for (filtering) food.[cxiii] In order that nothing rough or harsh reaches the liver from the stomach, (look) how they were created thin and fine so that they do not carry inferior (and harmful) materials (to the liver).[cxiv] (Ponder how) the liver changes these into blood by the permission of God ﷻ.[cxv]

This is then transported to the rest of the body through passageways specifically constructed for that purpose.[cxvi] So (observe) from that (blood): what reaches every part of the body is what is most appropriate for it, whether it is dry or soft (nutrients) or anything else. "*So Glorified is God, the Lord of the Worlds.*"[cxvii] Harmful and excessive (waste) arrive at receptacles[cxviii] and limbs that have been made for that (purpose of excretion), as we have previously mentioned. Being like vessels, they carry these excess waste products in order that they do not disperse in the rest of the body and harm it.

Then look and see if you find anything in the creation of the body that has no purpose. Was eyesight created for any other purpose except for him to perceive things and colours? If there were colours but no eyesight to perceive them, would there be any benefit in the existence of such colours? If there was no light emanating besides the inherent light of the eyes, what benefit would be gained from eyesight? Was the faculty of hearing created for any other purpose except for him to perceive different voices? If there were sounds but hearing did not exist to perceive them, would there be any benefit in the existence of such sounds? Likewise for all the senses.

الْفَضَلَاتِ. وَإِذَا تَأَمَّلْتَ كَذَلِكَ مَعَ سَائِرِ مَا فِي الْإِنْسَانِ وَجَدْتَهُ قَدْ وُضِعَ عَلَى غَايَةِ الْحِكْمَةِ وَالصَّوَابِ. فَكِّرْ فِي وُصُوْلِ الْغِذَاءِ إِلَى الْمَعِدَةِ حَتَّى تُنْضِجَهُ، وَتَبْعَثَ صَفْوَهُ إِلَى الْكَبِدِ فِي عُرُوْقٍ دِقَاقٍ قَدْ جُعِلَتْ كَالْمِصْفَاةِ لِلْغِذَاءِ، وَلِكَيْلَا يَصِلَ إِلَى الْكَبِدِ مِنْهُ شَيْءٌ غَلِيْظٌ خَشِنٌ فَيَنْكَؤُهَا، فَإِنَّهَا خُلِقَتْ دَقِيْقَةً لَا تَحْمِلُ الْغَثَّ، فَتَقْلِبُهُ بِإِذْنِ اللهِ دَماً، وَتَنْفُذُ بِهِ إِلَى سَائِرِ الْبَدَنِ فِي مَجَارٍ مُهَيَّأَةٍ لِذَلِكَ، فَيَصِلُ إِلَى كُلِّ شَيْءٍ مِنْ ذَلِكَ مَا يُنَاسِبُهُ مِنْ يَابِسٍ وَرِخْوٍ وَغَيْرِ ذَلِكَ {فَتَبَارَكَ اللَّهُ رَبُّ الْعَالَمِينَ} [غافر: ٦٤]. ثُمَّ يَنْفُذُ مَا يَكُوْنُ مِنْ خَبَثٍ وَفُضُوْلٍ إِلَى مغايض وَأَعْضَاءٍ أُعِدَّتْ لِذَلِكَ كَمَا ذَكَرْنَا قَبْلَ هَذَا، فَكَوَّنَهَا كَالْأَوْعِيَةِ لِتَحْمِلَ هَذِهِ الْفَضَلَاتِ؛ لِكَيْلَا تَنْتَشِرَ فِي الْبَدَنِ فَتُقَسِّمَهُ. ثُمَّ انْظُرْ هَلْ تَجِدُ فِي خَلْقِ الْبَدَنِ شَيْئًا لَا مَعْنَى لَهُ؟ هَلْ خُلِقَ الْبَصَرُ إِلَّا لِيُدْرِكَ الْأَشْيَاءَ وَالْأَلْوَانَ؟ فَلَوْ كَانَتِ الْأَلْوَانُ وَلَمْ يَكُنْ بَصَرٌ يُدْرِكُهَا، هَلْ كَانَ فِي الْأَلْوَانِ مَنْفَعَةٌ؟ وَلَوْ لَمْ يَكُنْ لِخَلْقِ الْأَبْصَارِ نُوْرٌ خَارِجٌ عَنْ نُوْرِهَا مَا كَانَ يُنْتَفَعُ بِالْبَصَرِ. وَهَلْ خُلِقَ السَّمْعُ إِلاَّ لِيُدْرِكَ الْأَصْوَاتَ؟ فَلَوْ كَانَتِ الْأَصْوَاتُ وَلَمْ يَكُنْ سَمْعٌ يُدْرِكُهَا لَمْ يَكُنْ فِي الْأَصْوَاتِ مَنْفَعَةٌ، وَكَذَلِكَ سَائِرُ الْحَوَاسِّ.

Ponder on the mediums that were created between the senses and perceivable things; the senses would be incomplete except through these mediums, which include light and air.[cxix] Had there not been light to make all visible things evident, then eyesight alone would not be able to convey (information about them). Had there not been air to carry sounds to the ears, (having) the ability to hear would not achieve (the complete sense of) hearing.

Ponder upon the one who lacks eyesight and hearing, and the afflictions he suffers from: indeed, he cannot see where he places his foot, cannot know what is in front of him, cannot distinguish between (various) colours, cannot sense a violent attack or an enemy (and protect himself), and cannot learn most crafts and trades. As for the one who lacks hearing, indeed he is one who is deprived of the sounds of conversations and meetings, and he is deprived of the pleasantness (found) in beautiful voices and harmonious melodies. He becomes burdensome upon the one who (attempts) to speak to him, such that he (the speaker) leaves him, and he is not able to hear anything from the news of people and their conversations. He becomes like the one who is absent (from a conversation) even though he is present, and like a dead person even though he is alive. As for the one who lacks intelligence, (look how) he becomes worse than the beasts.

Then look at how these limbs came to be, and their descriptions. These limbs (and faculties) bring about his wholesomeness, are sufficient for all his needs, and are complementary (with one another) towards his aims and purposes. If he were devoid of any (of these limbs), he would be found wanting, and his state would be difficult. Whoever is tested by lacking one of these things, then (understand) that is an education, exhortation, and recognition (for man to acknowledge) the true extent of the favour of God in its proper right and the right of other similar (blessings). Through

فَكِّرْ فِي أَشْيَاءَ جُعِلَتْ بَيْنَ الْحَوَاسِّ وَالْمَحْسُوْسَاتِ لَا يَتِمُّ الْحِسُّ إِلَّا بِهَا، مِنْهَا الضِّيَاءُ وَالْهَوَاءُ، فَلَوْ لَمْ يَكُنْ ضِيَاءٌ تَظْهَرُ فِيْهِ الْمُبْصَرَاتُ لَمْ يُدْرِكْهَا الْبَصَرُ، وَلَوْ لَمْ يَكُنْ هَوَاءٌ يُؤَدِّي الصَّوْتَ إِلَى السَّمْعِ لَمْ يَكُنِ السَّمْعُ يُدْرِكُ الصَّوْتَ.

فَكِّرْ فِيْمَنْ عَدِمَ الْبَصَرَ وَالسَّمْعَ وَمَا يَنَالُهُ مِنَ الْخَلَلِ، فَإِنَّهُ لَا يَنْظُرُ أَيْنَ يَضَعُ قَدَمَهُ، وَلَا يَدْرِيْ مَا بَيْنَ يَدَيْهِ، وَلَا يُفَرِّقُ مَا بَيْنَ الْأَلْوَانِ، وَلَا يَدْرِيْ بِهُجُوْمِ آفَةٍ أَوْ عَدُوٍّ، وَلَا سَبِيلَ لَهُ أَنْ يَتَعَلَّمَ أَكْثَرَ الصِّنَاعَاتِ، وَأَمَّا مَنْ عَدِمَ السَّمْعَ فَإِنَّهُ يَفْقِدُ رُوْحَ الْمُخَاطَبَةِ وَالْمُحَاضَرَةِ، وَيَعْدِمُ لَذَّةَ الْأَصْوَاتِ الْمُسْتَحْسَنَةِ وَالْأَلْحَانِ الْمُطْرِبَةِ، وَتَعْظُمُ الْمُؤْنَةُ عَلَى مَنْ يُخَاطِبُهُ حَتَّى يَنْصَرِمَ مِنْهُ، وَلَا يَسْمَعُ شَيْئاً مِنْ أَخْبَارِ النَّاسِ وَأَحَادِيْثِهِمْ حَتَّى يَصِيْرَ كَالْغَائِبِ وَهُوَ شَاهِدٌ، وَكَالْمَيِّتِ وَهُوَ حَيٌّ، وَأَمَّا مَنْ عَدِمَ الْعَقْلَ فَهُوَ أَشَرُّ مِنَ الْبَهَائِمِ.

فَانْظُرْ كَيْفَ صَارَتْ هَذِهِ الْجَوَارِحُ وَهَذِهِ الْأَوْصَافُ الَّتِي بِهَا صَلَاحُ الْإِنْسَانِ مُحَصِّلَةً وَمُبَلِّغَةً لِجَمِيعِ مَآرِبِهِ، وَمُتَمِّمَةً لِجَمِيعِ مَقَاصِدِهِ، وَإِذَا فَقَدَ شَيْئاً اخْتَلَّ أَمْرُهُ وَعَظُمَ مُصَابُهُ، وَمَنْ بُلِيَ بِفَقْدِ شَيْءٍ مِنْهَا فَهُوَ تَأْدِيْبٌ وَمَوْعِظَةٌ وَتَعْرِيْفٌ بِقَدْرِ نِعْمَةِ اللهِ فِي حَقِّهِ وَحَقِّ أَمْثَالِهِ، وَلِيَنَالَ

his patience, surely a better allotment will reach him in the Hereafter.[cxx,cxxi] So look at how the mercy of God can be found in both giving and withholding![cxxii]

Then ponder upon the limbs – those that have been created singly and in pairs – and consider what Divine Wisdom and propriety is contained therein. The head is a limb that is created singly, and despite the existence of many senses, one head is capable of containing all of them. (Understand that) if anything were added to it, then it would become heavy, and he would not have any need of it. If there were two (heads), one would become inoperative when the other speaks, and there would not be any need of it. If both of them were able to speak with one voice, then one of them would become excessive, and there would be no need for it. But if both of them were able to speak differently than what the other spoke, then the listener would never comprehend the purpose of such speech, for the listener is only able to grasp that which is intelligible and comprehensible.[cxxiii]

The hands were created in pairs, and (understand) that there would be no goodness for man if he were made to suffer with only one hand due to (his inability to perform) various tasks that concern him. Indeed, you can see the (amount of) inability one must bear when paralyzed in one hand. If he were entrusted with something, he would not be able to carry it out, nor would he be able to accomplish what one having two hands could accomplish. As for the Divine Wisdom in two feet, this (Wisdom) is clearly evident.[cxxiv]

Ponder upon the arrangement of the instruments of voices: the larynx is like a pipe for the exiting of the voice, while the tongue, lips, teeth, and mouth are for the forming of letters. Do you not see the example of one whose teeth have either (completely) fallen out or mostly (fallen out): what sorts of blemishes occur in his speech? Then look at the throat and the blessings and benefits contained

بِصَبْرِهِ عَلَى ذَلِكَ حَظّاً فِي الْآخِرَةِ، فَانْظُرْ إِلَى رَحْمَةِ اللهِ كَيْفَ تُوْجَدُ فِي الْعَطَاءِ والْمَنْعِ.

ثُمَّ فَكِّرْ فِي الْأَعْضَاءِ الَّتِي خُلِقَتْ أَفْرَاداً وَأَزْوَاجاً، وَمَا فِي ذَلِكَ مِنَ الْحِكْمَةِ وَالصَّوَابِ، فَالرَّأْسُ مِمَّا خُلِقَ فَرْداً، وَإِنَّ كَثِيْراً مِنَ الْحَوَاسِّ قَدْ حَوَاهَا رَأْسٌ وَاحِدٌ، وَلَوْ زَادَ عَلَيْهِ شَيْءٌ كَانَ ثَقِيْلاً لَا يُحْتَاجُ إِلَيْهِ، فَإِنْ كَانَ قِسْمَيْنِ فَإِنْ تَكَلَّمَ وَاحِدُهُمَا بَقِيَ الْآخَرُ مُعَطَّلاً لَا حَاجَةَ إِلَيْهِ، وَإِنْ تَكَلَّمَ مِنْهُمَا جَمِيْعاً بِكَلَامٍ وَاحِدٍ كَانَ أَحَدُهُمَا فَضْلَةً لَا يُحْتَاجُ إِلَيْهَا، وَإِنْ تَكَلَّمَ مِنْ أَحَدِهِمَا بِخِلَافِ مَا يَتَكَلَّمُ بِهِ مِنْ الْآخَرِ لَمْ يَدْرِ السَّامِعُ مُرَادَهُ مِنْ ذَلِكَ، وَأَمَّا الَّذِي يَأْخُذُ بِهِ السَّامِعُ هُوَ مَا كَانَ وَاضِحاً. وَالْيَدَانِ خُلِقَتَا أَزْوَاجاً، وَلَوْ لَمْ يَكُنْ لِلْإِنْسَانِ خَيْرٌ فِي أَنْ يَكُوْنَ يُلِمُّ بِيَدٍ وَاحِدَةٍ لَاخْتَلَّ مَا يُعَالِجُهُ مِنَ الْأُمُوْرِ، فَإِنَّكَ تَرَى مَنْ شُلَّتْ إِحْدَى يَدَيْهِ مَا يَكُوْنُ عِنْدَهُ مِنَ النَّقْصِ، وَإِنْ يُكَلَّفْ بِشَيْءٍ لَمْ يُحْكِمْهُ، وَلَا يَبْلُغُ فِيْهِ مَا يَبْلُغُ صَاحِبُ الْيَدَيْنِ، وَحِكْمَةُ الرِّجْلَيْنِ ظَاهِرَةٌ.

فَكِّرْ فِي تَهْيِئَةِ آلَاتِ الصَّوْتِ، فَالحَنْجَرَةُ كَالْأُنْبُوْبَةِ لِخُرُوْجِ الصَّوْتِ، وَاللِّسَانُ، وَالشَّفَتَانِ، وَالْأَسْنَانُ لِإِصَاغَةِ الْحُرُوْفِ، وَالْفَمُ، أَلَا تَرَى أَنَّ مَنْ سَقَطَتْ أَسْنَانُهُ أَوْ أَكْثَرُهَا كَيْفَ يَحْصُلُ الْخَلَلُ فِي كَلَامِهِ؟ ثُمَّ انْظُرْ

therein, and how it exists to transport wind to the lung so that man can cool his heart with this uninterrupted breath.

(Ponder upon what Divine Wisdom) is contained in the tongue; how it aids in the turning over of food and helping man, permitting the process of digesting food and drink. And (what Divine Wisdom) is contained in the teeth as they also aid the tongue, for they are like a support for the two lips as they hold back and release them from the front of the mouth. And (what Divine Wisdom) is contained in the two lips, for he is able to sip liquids, so that whatever enters his stomach does so with a purpose and an order that the human being voluntarily chooses. These lips also remain like a protective door over the mouth.[cxxv]

Thus, it should be evident to you that every organ among all of these organs is devoted to (accomplish) specific purposes and a variety of (bodily) functions needed for his well-being. (Consider) that if there were more (organs), his body would become spoiled, and if there were less, his body would become spoiled (as well). That is the Decree of the Almighty ﷻ, the All-Knowing ﷻ.

Then ponder upon the brain: if it were uncovered from him, you would find it in a condition where some of it is wrapped and coiled over itself to protect it from physical harm.[cxxvi] The skull serves as a covering over it, and the hair is a curtain (over that) and also a beautification. Both (the skull and hair) cause whatever may (potentially) harm him – such as heat, cold, and other such things – to be far away from it (the brain). He (may He be Glorified and Exalted) made this fortification for the brain in order to teach man its importance, and that it is most worthy (of this recognition) as the source of all perception.[cxxvii]

Then look at how the heart was concealed in the chest, and how He clothed it with armor that is only a membrane. Yet He perfected it, then further protected this with the surrounding ribs,[cxxviii] and

إِلَى مَا فِي الْحَنْجَرَةِ مِنَ الْمَنْفَعَةِ لِسُلُوْكِ النَّسِيْمِ مِنْهَا إِلَى الرِّئَةِ، فَتَرُوْحُ عَلَى الْفُؤَادِ بِهَذَا النَّفَسِ الْمُتَتَابِعِ، وَمَا فِي اللِّسَانِ مِنْ تَقْلِيْبِ الطَّعَامِ وَإِعَانَتِهِ عَلَى تَسْوِيْغِ الطَّعَامِ وَالشَّرَابِ، وَمَا فِي الْأَسْنَانِ مِنَ الْمَعُوْنَةِ أَيْضاً، ثُمَّ هِيَ كَالْمَسْنَدِ لِلشَّفَتَيْنِ تُمْسِكُهُمَا وَتَدْعَمُهُمَا مِنْ دَاخِلِ الْفَمِ، وَبِالشَّفَتَيْنِ يُرْتَشَفُ الشَّرَابُ حَتَّى يَكُوْنَ مَا يَدْخُلُهُ إِلَى الْجَوْفِ بِقَصْدٍ وَبِقَدْرِ مَا يَخْتَارُهُ الْإِنْسَانُ، ثُمَّ هُمَا عَلَى الْفَمِ كَالْبَابِ. فَقَدْ تَبَيَّنَ لَكَ أَنَّ كُلَّ عُضْوٍ مِنْ هَذِهِ الْأَعْضَاءِ يَنْصَرِفُ إِلَى وُجُوْهٍ مِنَ الْمَآرِبِ، وَضُرُوْبٍ مِنَ الْمَصَالِحِ، إِنْ زَادَ أَفْسَدَ، وَإِنْ نَقَصَ أَفْسَدَ، فَذَلِكَ تَقْدِيْرُ الْعَزِيْزِ الْعَلِيْمِ. فَكِّرْ فِي الدِّمَاغِ إِذَا كُشِفَ عَنْهُ فَإِنَّكَ تَجِدُهُ قَدْ لُفَّ بَعْضُهُ فَوْقَ بَعْضٍ؛ لِيَصُوْنَهُ مِنَ الْأَعْرَاضِ، وَأُطْبِقَتْ عَلَيْهِ الْجُمْجُمَةُ، وَالشَّعْرُ سَتْرٌ لَهَا وَجَمَالٌ، وَلِتُبْعِدَ عَنْهَا مَا يُؤْذِيْهَا مِنْ حَرٍّ وَبَرْدٍ وَغَيْرِ ذَلِكَ، فَحَصَّنَ سُبْحَانَهُ وَتَعَالَى الدِّمَاغَ هَذَا التَّحْصِيْنَ لِعِلْمِهِ بِأَنَّهُ مُهِمٌّ، وَأَنَّهُ مُسْتَحِقٌّ لِذَلِكَ؛ لِكَوْنِهِ يَنْبُوْعَ الْحِسِّ.

ثُمَّ انْظُرْ كَيْفَ غَيَّبَ الْفُؤَادَ فِي جَوْفِ الصَّدْرِ، وَكَسَاهُ الْمَدْرَعَةَ الَّتِي هِيَ غِشَاؤُهُ وَأَتْقَنَهَا، وَحَصَّنَهُ بِالْجَوَانِحِ وَمَا عَلَيْهَا مِنَ اللَّحْمِ وَالْعَصَبِ لِشَرَفِهِ، وَأَنَّ ذَلِكَ هُوَ اللَّائِقُ بِهِ.

whatever flesh and nerves are over it (the heart) are only for further honoring him. That is most suitable for this honoring.

Then look at how He made two passageways in the throat: one for speaking, and that is the part of the throat (i.e., the trachea) connected to the lung; and the other (passageway) for nourishment, and that is the esophagus connected to the stomach. (Then look how) He made a covering over the trachea (i.e., the epiglottis) that prevents food from reaching it.

Then He made the lungs as an apparatus for the aeration of the heart, which neither stop nor falter. They take (blood from the heart) and return it without any difficulty so that heat does not cause confinement and damage to the heart.[cxxix] He filled the atmosphere with air for this benefit and more.[cxxx,cxxxi]

Then look at how He made passageways for the excretion of urine and feces in order to contain (and regulate) them so that they do not flow out continuously and corrupt the wholesomeness of human beings.

Then look at the flesh and muscles of the two thighs as plentiful and thick. This is such that they protect man from the pain of sitting upon the earth. The person whose body is too delicate and scrawny suffers (pain) from sitting, as if there was nothing between him and the floor.

Then consider that if man's penis were perpetually flaccid, how would sperm ever arrive at the place of creation (i.e., the womb)? If his penis were perpetually erect, how would his state be in his conduct (and daily affairs) while it is in that state? Rather, He made his penis hidden as if passion was never created for him. Consider (this): is it not the most excellent of planning in creation that, just as there is a latrine in the most hidden of places in the house, so (too) for this purpose there is an orifice (i.e., the anus) prepared for answering the call of nature located in the most concealed place of

ثُمَّ انْظُرْ كَيْفَ جَعَلَ فِي الْحَلْقِ مَنْفَذَيْنِ: أَحَدُهُمَا لِلصَّوْتِ وَهُوَ الْحُلْقُومُ الْوَاصِلُ إِلَى الرِّئَةِ، وَالْآخَرُ لِلْغِذَاءِ وَهُوَ الْمَرِيءُ الْوَاصِلُ إِلَى الْمَعِدَةِ، وَجَعَلَ عَلَى الْحُلْقُومِ طَبَقاً يَمْنَعُ الطَّعَامَ أَنْ يَصِلَ إِلَيْهِ، ثُمَّ جَعَلَ الرِّئَةَ مِرْوَحَةَ الْفُؤَادِ لَا تَفْتُرُ وَلَا تُخِلُّ، تَأْخُذُ وَتَرُدُّ بِغَيْرِ كُلْفَةٍ؛ لِئَلَّا تَنْحَصِرَ الحَرَارَةُ فِي الْقَلْبِ فَتُؤَدِّيَ إِلَى التَّلَفِ، ثُمّ مَلَأَ الجَوَّ هَوَاءً لِهَذِهِ الْمَصْلَحَةِ وَلِغَيْرِهَا. ثُمَّ انْظُرْ كَيْفَ جَعَلَ لِمَنَافِذِ الْبَوْلِ وَالْغَائِطِ سَرَاحاً يَضْبِطُهَا؛ لِكَيْ لَا يَجْرِيَ جَرَيَاناً دَائِماً فَيُفْسِدَ عَلَى الْإِنْسَانِ عَيْشَهُ، ثُمَّ انْظُرْ كَيْفَ جَعَلَ لَحْمَ الْفَخِذَيْنِ كَثِيْراً كَثِيْفاً لِيَقِيَ الْإِنْسَانَ مِنْ أَلَمِ الْجُلُوسِ عَلَى الْأَرْضِ، كَمَا يَأْلَمُ مِنَ الْجُلُوْسِ مَنْ نَحُلَ جِسْمُهُ وَقَلَّ لَحْمُهُ إِذَا لَمْ يَكُنْ بَيْنَهُ وَبَيْنَ الْأَرْضِ حَائِلٌ. اُنْظُرْ لَوْ كَانَ ذَكَرُ الْإِنْسَانِ مُسْتَرْخِياً أَبَداً كَيْفَ يَصِلُ الْمَاءُ إِلَى مَوْضِعِ الْخَلْقِ، وَلَوْ كَانَ مُنْعَظاً أَبَداً كَيْفَ يَكُوْنُ حَالُهُ فِي تَصَرُّفَاتِهِ وَهُوَ كَذَلِكَ؟ بَلْ جَعَلَهُ مَسْتُوْراً كَأَنَّهُ لَمْ تُخْلَقْ لَهُ شَهْوَةٌ. ثُمَّ انْظُرْ أَلَيْسَ أَنَّهُ مِنْ حُسْنِ التَّدْبِيْرِ فِي الْبِنَاءِ أَنْ يَكُوْنَ الْخَلَاءُ فِي أَسْتَرِ مَوْضِعٍ فِي الدَّارِ؟ فَلِهَذَا اتُّخِذَ الْمَنْفَذُ الْمُهَيَّأُ لِقَضَاءِ حَاجَةِ الْإِنْسَانِ فِي أَسْتَرِ مَوْضِعٍ مِنْ جَسَدِهِ مَغَيَّبٍ فِيْهِ، تَلْتَقِي عَلَيْهِ فَخِذَاهُ بِمَا عَلَيْهِمَا مِنَ اللَّحْمِ فَتُوَارِيْهِ بِهِ، وَيَخْفَى ذَكَرُهُ،

his body, (itself) hidden to him; the fleshiness of his buttocks covers it so that it is not easily noticeable. This is a specific (design) for man to honor him.[cxxxii]

Then look at the creation of hair and nails: whenever they become long, there is a wholesomeness in their trimming.[cxxxiii] They were created devoid of (pain) sensation, such that pain does not reach man when they are being cut. If it were not for this Divine Wisdom, then one of two matters would occur: either he would leave them (his hair and nails) in their natural condition (to grow without restriction), causing him to become ugly, or he would (completely) remove them but would be harmed by this removal.

Then ponder upon man's hair. If it originated in the eye, then his vision would be blinded; or if it was in the mouth, it would disturb his eating and drinking; or if it was in the palm of his hand, it would ruin the joy of touch and (his ability to perform) certain actions; or if it was in the private parts, it would spoil the joy of intimate relations that result from the meeting of these two places despite its ability to grow (there). So Glorified is the Planner and Bestower of this blessing!

Look at how creation has been intended to follow the right path and be protected from harm and injury. Then (look at) what natural needs have been put into man in terms of food, sleep, and intimate relations, and what perfect planning lies therein. Indeed, He created an impulse in man's inner disposition to fulfill and incite him towards (completing) these needs. Hunger and thirst require him to seek food, through which his life is preserved, and likewise is (his seeking) drink, through which he is supported. In sleep, there is rest for the body and a renewal of strength, whereas lust necessitates intimate relations, only through which lineage is maintained and mankind's remaining (on the Earth) is preserved.

If man was to eat only because of his knowledge of its necessity,

وَذَلِكَ مَخْصُوْصٌ بِالْإِنْسَانِ لِشَرَفِهِ. ثُمَّ انْظُرْ فِي خَلْقِ الشَّعْرِ وَالْأَظْفَارِ لَمَّا كَانَ يَطُوْلَانِ، وَفِي تَقْصِيْرِهِمَا مَصْلَحَةٌ جُعِلَا عَدِيْمَي الْحِسِّ؛ حَتَّى لَا يَنَالَ الْإِنْسَانَ أَلَمٌ عِنْدَ التَّزَيُّنِ بِقَصِّهِمَا، وَلَوْلَا هَذِهِ الْحِكْمَةُ لَكَانَ بَيْنَ أَمْرَيْنِ: إِمَّا أَنْ يَدَعَهُمَا عَلَى حَالِهِمَا فَيَتَشَوَّهَ خَلْقُهُ، أَوْ يُزِيْلَ ذَلِكَ فَيَتَأَلَّمَ بِإِزَالَتِهِ.

ثُمَّ تَفَكَّرْ فِي الشُّعُوْرِ لَوْ نَبَتَتْ فِي الْعَيْنِ لَأَعْمَتِ الْبَصَرَ، أَوْ فِي الْفَمِ لَنَغَّصَتِ الْأَكْلَ وَالشُّرْبَ، أَوْ فِي رَاحَةِ الْكَفِّ لَنَفِدَتْ لَذَّةُ اللَّمْسِ وَبَعْضُ الْأَعْمَالِ، أَوْ فِي الْفَرْجِ لَكَدَّرَتْ لَذَّةَ الْجِمَاعِ، مَعَ قَبُولِ هَذِهِ الْمَوَاضِعِ لِنَبَاتِهَا فِيْهَا، فَسُبْحَانَ الْمُدَبِّرِ الْمُنْعِمِ بِهَذِهِ النِّعَمِ.

فَانْظُرْ كَيْفَ قُصِدَ بِهَذَا الْخَلْقِ طَرِيْقُ الصَّوَابِ، وَتَجَنُّبُ الْخَطَأِ وَالضَّرَرِ، ثُمَّ فِيْمَا جُبِلَ عَلَيْهِ الْإِنْسَانُ مِنَ الِاحْتِيَاجِ إِلَى الْمَطْعَمِ وَالنَّوْمِ وَالْجِمَاعِ، وَمَا فِي ذَلِكَ مِنَ التَّدْبِيْرِ الْمُحْكَمِ، فَقَدْ جَعَلَ فِي طَبْعِهِ مُحَرِّكاً يَقْتَضِيْهِ وَيَسْتَحِثُّهُ، فَالْجُوْعُ وَالْعَطَشُ يَقْتَضِيْ طَلَبَ الطَّعَامِ الَّذِي بِهِ حَيَاتُهُ، وَكَذَلِكَ الشَّرَابُ الَّذِي بِهِ قِوَامُهُ، وَالنَّوْمُ فِيْهِ رَاحَةٌ لِلْبَدَنِ وَعُمُوْمِ الْقُوَى، وَالشَّبَقُ يَقْتَضِيْ الْجِمَاعَ الَّذِي بِهِ دَوَامُ النَّسْلِ وَبَقَاؤُهُ، فَلَوْ كَانَ الْإِنْسَانُ إِنَّمَا يَتَنَاوَلُ الطَّعَامَ وَالشَّرَابَ لِمَعْرِفَتِهِ بِالْحَاجَةِ إِلَيْهِ،

and had no natural inclination towards it, he would be (easily) preoccupied with pursuing his daily affairs until his strength melts away and he dies. (This is like) if one has a need for some medicine but dislikes it despite its wholesomeness. If a pleasure and need for it is not in his natural disposition, then (this aversion) prevents it from reaching him, so he will remain ill or even perish.[cxxxiv] Likewise, if he did this with sleep, and the desire to sleep only entered his body (i.e., his consciousness) according to his choice for it, then he would be distracted and preoccupied by various tasks (and never choose to sleep), and his body would die from tiredness and exhaustion. Likewise is his undaunted desire towards intimate relations: had the desire to acquire children not been in him, then indeed all lineage would have been cut off due to any difficult obstacle that worked against him (to achieve this).[cxxxv] So look how an inherent disposition has been made in him (to endure) that which harms him in order to attain these (aforementioned) benefits.

Then look at how these (four) faculties (i.e., eating, drinking, sleeping, and mating) were put into proper order by virtue of this magnificent plan. The entire body is like the position of a king's house: there is a retinue and servants (that work together) entrusted with the ordered running of the house; some work for the needs of the retinue and bring water to them; others assigned for the task of keeping what needs to be stored until it is treated and prepared; others for the well-being and preparedness of it – its wellbeing, which I have specified previously; others assigned to sweep whatever filth and waste that may be in the house.[cxxxvi] The king in this example is the All-Knowing Creator (may He be Glorified); the realm is the body; the retinue represents the limbs; and the servants are the four faculties, those which are the soul and its places of occurrence in man are in the meanings found in contemplation, imagination, intellect, watchfulness, anger, and other such things. Do you not

وَلَمْ يَجِدْ مِنْ طِبَاعِهِ مَا يُلْجِئُهُ إِلَيْهِ لَاشْتَغَلَ بِأَسْبَابِ ضَرُوْرَتِهِ فَتَنْحَلُّ قُوَاهُ وَيَهْلِكُ، كَمَا أَنَّهُ قَدْ يَحْتَاجُ إِلَى دَوَاءٍ يَكْرَهُهُ وَفِيْهِ صَلَاحُهُ، وَلَيْسَ فِي جِبلَّتِهِ دَاعِيَةٌ لَهُ فَيُدَافِعَ عَنْ تَنَاوُلِهِ فَيَمْرَضَ أَوْ يَمُوْتَ، فَكَذَلِكَ لَوْ كَانَ يَفْعَلُ النَّوْمَ وَيُدْخِلُهُ عَلَى جِسْمِهِ بِاخْتِيَارِهِ لَتَشَاغَلَ عَنْهُ بِبَعْضِ مُهِمَّاتِهِ فَيَهْلِكَ جِسْمُهُ بِالتَّعَبِ وَالنَّصَبِ، وَكَذَلِكَ لَوْ كَانَ إِقْدَامُهُ عَلَى الْجِمَاعِ إِنَّمَا هُوَ لِرَغْبَةِ حُصُوْلِ الْوَلَدِ لَانْقَطَعَ النَّسْلُ لِمَا يُعَارِضُهُ مِنَ الْأَسْبَابِ الْمُشْغِلَةِ، فَانْظُرْ كَيْفَ جُعِلَ فِيْهِ بِالطَّبْعِ مَا يَضْطَرُّهُ إِلَى حُصُوْلِ هَذِهِ الْفَوَائِدِ.

ثُمَّ انْظُرْ كَيْفَ رُتِّبَتْ هَذِهِ الْقُوَى بِهَذَا التَّرْتِيبِ الْمُحْكَمِ الْعَجِيْبِ، فَصَارَ الْبَدَنُ بِمَا فِيْهِ بِمَنْزِلَةِ دَارٍ لِمَلِكٍ فِيْهَا حَشَمٌ، وَقَوْمٌ مُوَكَّلُوْنَ بِالدَّارِ، فَوَاحِدٌ لِإِمْضَاءِ حَوَائِجِ الْحَشَمِ وَإِيْرَادِ مَاءٍ لَهُمْ، وَآخَرُ لِقَبْضِ مَا يُرَادُ خَزْنُهُ إِلَى أَنْ يُعَالَجَ وَيُهَيَّأَ، وَآخَرُ لِإِصْلَاحِ ذَلِكَ وَتَهْيِئَتِهِ، وَإِصْلَاحُهُ أَخَصُّ مِمَّا قَبْلُ، وَآخَرُ لِكَسْحِ مَا فِي الدَّارِ مِنَ الْأَقْذَارِ وَإِخْرَاجِهِ، فَالْمَلِكُ فِي هَذَا الْمَثَلِ هُوَ الْخَالِقُ الْعَلِيْمُ سُبْحَانَهُ، وَالدَّارُ هِيَ الْبَدَنُ، وَالْحَشَمُ هِيَ الْأَعْضَاءُ، وَالْقَوْمُ فِي هَذِهِ الْقُوَى الْأَرْبَعِ هِيَ النَّفْسُ، وَمَوْقِعُهَا مِنَ الْإِنْسَانِ بِمَعْنَى الْفِكْرِ وَالْوَهْمِ وَالْعَقْلِ وَالْحِفْظِ

see how man's condition would be if even one of those protective attributes were lacking? He would not be able to remember that which is his (nor protect against) that which is against him; (nor have knowledge of) what he has sent forth or what is coming to him; what he has given or what he took; what he saw and what he heard; what he said or what was said to him; he would not be able to remember who has done good to him and who has done evil to him; and who has benefited him from who has harmed him. He would not willfully accomplish (a task) even if he had purposely intended it, nor (attain) knowledge even if he studied it, and would not derive any benefit from his freedom (of will). He would not be able to draw any lesson from what has transpired in the past.

Then look at these blessings: how (great) is the importance of each one of them, so how (great) are all of them? Yet even more wondrous than the blessing of remembering is the blessing of forgetting. Had it not been for forgetfulness, man would not be able to withdraw himself from difficulties or alleviate his regret. Rancor would never leave him, and he would not be able to enjoy any worldly and sensory pleasures due to remembering calamities and frustrating circumstances.[cxxxvii] He would be incapable of expecting forgetfulness from an oppressor or (giving) respite, or distractedness from an envier or someone who intends him harm.[cxxxviii] So look how God (may He be Glorified) made in him both (faculties of) watchfulness and forgetfulness, even though these two (faculties) are completely opposite; yet He made in each of these two faculties certain kinds of benefit for man.

Then look towards what He exclusively bestowed upon him concerning modesty amongst all animals. Had it not been for this (modesty), his shortcomings would never be forgiven, his requests would never be answered, the guest would never be honored, good deeds would never be appreciated and reciprocated, and evil deeds

وَالْغَضَبِ وَغَيْرِ ذَلِكَ. أَرَأَيْتَ لَوْ نَقَصَ مِنَ الْإِنْسَانِ مِنْ هَذِهِ الصِّفَاتِ الْحِفْظُ وَحْدَهُ كَيْفَ يَكُوْنُ حَالُهُ؟ وَكَانَ لَا يَحْفَظُ مَا لَهُ وَمَا عَلَيْهِ، وَمَا أَصْدَرَ وَمَا أَوْرَدَ، وَمَا أَعْطَى وَمَا أَخَذَ، وَمَا رَأَى وَمَا سَمِعَ، وَمَا قَالَ وَمَا قِيلَ لَهُ، وَلَمْ يَذْكُرْ مَنْ أَحْسَنَ إِلَيْهِ وَلَا مَنْ أَسَاءَ لَهُ، وَلَا مَنْ نَفَعَهُ مِمَّنْ ضَرَّهُ، وَكَانَ لَا يَهْتَدِيْ لِطَرِيْقٍ وَلَوْ سَلَكَهُ، وَلَا لِعِلْمٍ وَلَوْ دَرَسَهُ، وَلَا يَنْتَفِعُ بِتَحْرِيْرِهِ، وَلَا يَسْتَطِيعُ أَنْ يَعْتَبِرَ بِمَنْ مَضَى.

فَانْظُرْ إِلَى هَذِهِ النِّعَمِ كَيْفَ مَوْقِعُ الْوَاحِدَةِ مِنْهَا؟ فَكَيْفَ جَمِيْعُهَا؟ وَأَعْجَبُ مِنْ نِعْمَةِ الْحِفْظِ نِعْمَةُ النِّسْيَانِ، فَلَوْلَا النِّسْيَانُ مَا سَلَا الْإِنْسَانُ عَنْ مُصِيْبَتِهِ، فَكَانَ لَا يَنْقُصُ لَهُ حَسْرَةٌ، وَلَا يَذْهَبُ عَنْهُ حِقْدٌ، وَلَا يَسْتَمْتِعُ بِشَيْءٍ مِنْ لَذَّاتِ الشَّهَوَاتِ الدُّنْيَوِيَّةِ مَعَ تَذَكُّرِ الْآفَاتِ وَالْفَجَائِعِ الْمُغْضِبَاتِ، وَكَانَ لَا يُمْكِنُ أَنْ يَتَوَقَّعَ غَفْلَةً مِنْ ظَالِمٍ، وَلَا فَتْرَةً وَلَا ذُهُوْلاً مِنْ حَاسِدٍ أَوْ قَاصِدِ مَضَرَّةٍ، فَانْظُرْ كَيْفَ جَعَلَ اللهُ فِيْهِ - سُبْحَانَهُ - الْحِفْظَ وَالنِّسْيَانَ وَهُمَا مُتَضَادَّانِ، وَجَعَلَ لِلْإِنْسَانِ فِي كُلٍّ مِنْهُمَا ضُرُوْباً مِنَ الْمَصَالِحِ.

ثُمَّ انْظُرْ إِلَى مَا خَصَّهُ بِهِ دُوْنَ غَيْرِهِ مِنَ الْحَيَوَانِ مِنَ الْحَيَاءِ، فَلَوْلَاهُ لَمْ تَقُلِ الْعَثَرَاتُ، وَلَمْ تُقْضَ الْحَاجَاتُ، وَلَمْ يُقْرَ الضَّيْفُ، وَلَمْ يُثْمِرِ

would never be abhorred and abandoned. In reality, many obligatory matters are done as a result of modesty (and fear of shame) among mankind.[cxxxix] (This is why) trusted things are returned, the rights of parents are protected, and other such things. He is made to refrain from committing illicit sexual relations and other such things due to (his desire to maintain social) propriety. Look at the grandest status of this blessing[cxl] according to this description.[cxli]

Then look at what God favored him with through the faculty of speech, which distinguishes him from beasts. He is able to express clearly what is contained in himself (i.e., his thoughts) and can comprehend from others what is (similarly) in their selves.[cxlii]

Likewise, the blessing of writing, through which the stories of past generations may benefit present generations, and the stories of the present may benefit those yet to come.[cxliii] Through (the faculty of writing) the sciences of knowledge and propriety are preserved in books. People may learn what happens among them from records of calculations and business transactions. Had it not been for writing, the news of some ages would be cut off from (reaching) other (ages), and though (some) knowledge would be studied, (knowledge of) past virtues and propriety would vanish. Since such knowledge would not exist, defects of character would become many and apparent.

If you said: speech and writing are faculties earned by mankind and are not matters created inherently in his nature, and because of that, the scripts of the Arabs, Indians, Greeks, and other peoples are different from one another; so likewise, speech is something specific (for a given group of people), and hence there are so many differences.[cxliv,cxlv] We reply: (understand that) those faculties through which writing is accomplished, such as the hand, fingers, and palm – all of which are proper for writing; the mind and the ability to ponder – all guide him such that he is able to write, but

الْجَمِيْلُ فَيُفْعَلَ، وَلَا يُتَجَافَى عَنِ الْقَبِيْحِ فَيُتْرَكَ، حَتَّى إِنَّ كَثِيْراً مِنَ الْأُمُوْرِ الْوَاجِبَةِ إِنَّمَا تُفْعَلُ لِسَبَبِ الْحَيَاءِ مِنَ النَّاسِ، فَتُرَدُّ الْأَمَانَاتُ، وَتُرَاعَى حُقُوْقُ الْوَالِدَيْنِ وَغَيْرِهِمَا، وَيُعَفُّ عَنْ فِعْلِ الْفَوَاحِشِ إِلَى غَيْرِ ذَلِكَ مِنْ أَجْلِ الْحَيَاءِ، فَانْظُرْ مَا أَعْظَمَ مَوْقِعَ هَذِهِ النِّعْمَةِ فِي هَذِهِ الصِّفَةِ.

وَانْظُرْ مَا أَنْعَمَ اللّٰهُ بِهِ مِنَ النُّطْقِ الَّذِي يُمَيَّزُ بِهِ عَنْهُ الْبَهَائِمُ، فَيُعَبِّرُ بِمَا فِي ضَمِيْرِهِ، وَيَفْهَمُ عَنْ غَيْرِهِ مَا فِي نَفْسِهِ، وَكَذَلِكَ نِعْمَةُ الْكِتَابَةِ الَّتِي تُفِيْدُ أَخْبَارَ الْمَاضِيْنَ لِلْبَاقِيْنَ، وَأَخْبَارَ الْبَاقِيْنَ لِلْآتِيْنَ، وَبِهَا تُخَلَّدُ فِي الْكُتُبِ الْعُلُوْمُ وَالْآدَابُ، وَيُعَلَّمُ النَّاسُ ذِكْرَ مَا يَجْرِيْ بَيْنَهُمْ فِي الْحِسَابِ وَالْمُعَامَلَاتِ، وَلَوْلَا الْكِتَابَةُ لَانْقَطَعَتْ أَخْبَارُ بَعْضِ الْأَزْمِنَةِ عَنْ بَعْضٍ، وَدُرِسَتِ الْعُلُوْمُ وَضَاعَتِ الْفَضَائِلُ وَالْآدَابُ، وَعَظُمَ الْخَلَلُ الدَّاخِلُ عَلَى النَّاسِ فِي أُمُوْرِهِمْ بِسَبَبِ عَدَمِهَا.

فَإِنْ قُلْتَ: إِنَّ الْكَلَامَ وَالْكِتَابَةَ مُكْتَسَبَةٌ لِلْإِنْسَانِ، وَلَيْسَتْ بِأَمْرٍ طَبِيْعِيٍّ، وَلِذَلِكَ تَخْتَلِفُ الْخُطُوْطُ بَيْنَ عَرَبِيٍّ وَهِنْدِيٍّ وَرُومِيٍّ إِلَى غَيْرِ ذَلِكَ، وَكَذَلِكَ الْكَلَامُ هُوَ شَيْءٌ يُصْطَلَحُ عَلَيْهِ، فَلِذَلِكَ اخْتَلَفَ.

قُلْنَا: مَا بِهِ تَحْصُلُ الْكِتَابَةُ مِنَ الْيَدِ وَالْأَصَابِعِ وَالْكَفِّ الْمُهَيَّأَ لِلْكِتَابَةِ، وَالذِّهْنِ وَالْفِكْرِ الَّذِي يَهْتَدِيْ بِهِ لَيْسَ بِفِعْلِ الْإِنْسَانِ، وَلَوْلَا

none of these exist through man's doing. Had it not been for these (aforementioned) faculties, he would never be able to write. Glory be to the Bestower of that (i.e., both the tools and ability to write)! Likewise, had it not been for the tongue, the inherent ability to speak, and the complex mind (that is able to construct intelligible words and phrases), he would never be able to speak. Glory be to the Bestower of that ability to speak upon him!

Then look at the Divine Wisdom in (the emotion of) anger created in man. Through this, he is able to defend himself from that which may harm him. (Look at the Divine Wisdom) that He created in him in the faculty of jealousy, through which he rushes towards attaining something and accomplishing that which benefits him. But even these faculties are matters ordained with moderation between two extremes.[cxlvi] If he were to transgress the boundary between these two emotions (i.e., anger and jealousy), he would join the rank of the devils.[cxlvii] So instead, it is necessary that he limit his anger only to repel harm and limit his jealousy towards righteous envy (*ghibṭah*) – this is a desire for that which benefits him without any harm reaching others.[cxlviii]

Then look at that which he has been given and that which he has been prevented from, for there is goodness in this as well.[cxlix] From this (goodness) is the faculty of hope through which the world flourishes and lineage continues, such that the weak may inherit the beneficial things of civilization from the strong. Indeed, man was weak when he was first created; had it not been for him finding the (constructed) legacies of people who came (before him) and left, he would never have a place of shelter nor a tool to benefit from.[cl] Hope (for the future) is a means through which the works of the present generation may occur in order to benefit future generations, allowing them to continue to inherit (from preceding generations) until the Day of Judgment.

ذَلِكَ لَمْ يَكُنْ لِيَكْتُبَ أَبداً، فَسُبْحَانَ الْمُنْعِمِ عَلَيْهِ بِذَلِكَ، وَكَذَلِكَ لَوْلَا اللِّسَانُ وَالنُّطْقُ الطَّبِيْعِيُّ فِيْهِ وَالذِّهْنُ الْمُرَكَّبُ فِيْهِ لَمْ يَكُنْ لِيَتَكَلَّمَ أَبَداً، فَسُبْحَانَ الْمُنْعِمِ عَلَيْهِ بِذَلِكَ.

ثُمَّ انْظُرْ إِلَى حِكْمَةِ الْغَضَبِ الْمَخْلُوْقِ فِيْهِ، يَدْفَعُ عَنْ نَفْسِهِ بِهِ مَا يُؤْذِيْهَا، وَمَا خُلِقَ فِيْهِ مِنَ الْحَسَدِ، فَبِهِ يَسْعَى فِي جَلْبِ مَا يَنْتَفِعُ بِهِ، غَيْرَ أَنَّهُ مَأْمُوْرٌ بِالِاعْتِدَالِ فِي هَذَيْنِ الْأَمْرَيْنِ، فَإِنْ جَاوَزَ الْحَدَّ فِيْهِمَا الْتَحَقَ بِرُتْبَةِ الشَّيَاطِيْنِ، بَلْ يَجِبُ أَنْ يَقْتَصِرَ فِي حَالَةِ الْغَضَبِ عَلَى دَفْعِ الضَّرَرِ، وَفِي الْحَسَدِ عَلَى الْغِبْطَةِ؛ وَهِيَ إِرَادَةُ مَا يَنْفَعُهُ مِنْ غَيْرِ مَضَرَّةٍ تَلْحَقُ غَيْرَهُ.

ثُمَّ انْظُرْ مَا أُعْطِيَ وَمَا مُنِعَ مِمَّا فِيْهِ أَيْضاً صَلَاحُهُ، فَمِنْ ذَلِكَ الْأَمَلُ، فَبِسَبَبِهِ تُعْمَرُ الدُّنْيَا، وَيَدُوْمُ النَّسْلُ لِيَرِثَ الضُّعَفَاءُ عَنِ الْأَقْوِيَاءِ مَنَافِعَ الْعِمَارَةِ، فَإِنَّ الْإِنْسَانَ أَوَّلُ مَا يُخْلَقُ ضَعِيْفٌ، فَلَوْلَا أَنَّهُ يَجِدُ آثَارَ قَوْمٍ أَحَلُّوا وَعَمَّرُوا لَمْ يَكُنْ لَهُ مَحَلٌّ يَأْوِيْ إِلَيْهِ، وَلَا آلَةٌ يَنْتَفِعُ بِهَا، فَكَانَ الْأَمَلُ سَبَباً لِعَمَلِ الْحَاضِرِيْنَ مَا يَقَعُ بِهِ انْتِفَاعُ الْآتِيْنَ، وَهَكَذَا يُتَوَارَثُ إِلَى يَوْمِ الدِّيْنِ.

وَمُنِعَ الْإِنْسَانُ مِنْ عِلْمِ أَجَلِهِ وَمَبْلَغِ عُمُرِهِ لِمَصْلَحَةٍ، فَإِنَّهُ لَوْ

Man has (also) been prevented from knowing the time of his death and the extent of his life for a benefit. Had man known that the extent of his life would be short, he would not enjoy life, be happy enough to reproduce, construct buildings on the Earth, and other such things. If he knew that the extent of his life would be lengthy, he would become engrossed in sensory pleasures, transgress all boundaries, and plunge himself into destructive pursuits; even the preacher (*wāʿiẓ*) would be incapable of stopping him (through preaching) and restraining him from that which would bring about his ruin. Therefore, in his ignorance of the length of his lifespan, there is a goodness that allows the occurrence of fear through the anticipation of the violence of (sudden) death, and a goodness that brings about (in him) an initiative to rectify his actions before his demise.

Then look at what benefits him from the purity and pleasures (experienced) from various types of foods despite their differences, and the (pleasure experienced from) fruits despite the differences of their colours and splendor.[cli] (Look at what benefits man from) the means of transportation, so that he may ride them and derive their benefits, and (the types of) birds so that he may find pleasure by listening to them.[clii,cliii,cliv] (Look at the benefits) in gold, silver, and precious stones: he acquires and arrives at his intended purposes through them and finds them (suitable to accomplish) his tasks; and the (benefits of) pharmacology that he uses in order to preserve his health; and animals for his consumption and other affairs such as plowing, carrying loads, and other functions. And (the benefits of) flowers and other fragrant things so that he may enjoy their scent and benefit from them, and (the benefits of) various clothing (he wears) despite the differences in origins.[clv] All of this is the fruit of whatever intelligence and understanding was created in him. Then look at all the wonders God has prepared in him (as well).[clvi]

عَلِمَ مُدَّةَ حَيَاتِهِ وَكَانَتْ قَصِيْرَةً لَمْ تَهْنَأْ حَيَاتُهُ، وَلَمْ يَنْشَرِحْ لِوُجُوْدِ نَسْلٍ وَلَا لِعِمَارَةِ أَرْضٍ وَلَا لِغَيْرِ ذَلِكَ، وَلَو عَلِمَهَا وَكَانَتْ طَوِيْلَةً لَانْـهَمَكَ فِي الشَّهَوَاتِ، وَتَعَدَّى الْحُدُوْدَ وَاقْتَحَمَ الْمُهْلِكَاتِ، وَلَعَجَزَ الْوُعَّاظُ عَنْ إِيْقَافِهِ وَزَجْرِهِ عَمَّا يُؤَدِّيْهِ إِلَى إِتْلَافِهِ، فَكَانَ فِي جَهْلِهِ بِمُدَّةِ عُمُرِهِ مَصْلَحَةُ حُصُوْلِ الْخَوْفِ بِتَوَقُّعِ هُجُوْمِ الْمَوْتِ، وَمُبَادَرَةِ صَالِحِ الْأَعْمَالِ قَبْلَ الْفَوَاتِ.

ثُمَّ انْظُرْ إِلَى مَا يَنْتَفِعُ بِهِ مِمَّا فِيْهِ مَصَالِحُهُ وَمَلَاذُّهُ مِنْ أَصْنَافِ الْأَطْعِمَةِ عَلَى اخْتِلَافِ طُعُوْمِهَا، وَأَصْنَافِ الْفَوَاكِهِ مَعَ اخْتِلَافِ أَلْوَانِهَا وَبَهْجَتِهَا، وَأَصْنَافِ الْمَرَاكِبِ لِيَرْكَبَهَا وَيُحَصِّلَ مَنَافِعَهَا، وَطُيُوْرٍ يَلْتَذُّ بِسَمَاعِهَا، وَنُقُوْدٍ وَجَوَاهِرَ يَقْتَنِيْهَا وَيَصِلُ بِهَا إِلَى أَغْرَاضِهِ وَيَجِدُهَا فِي مُهِمَّاتِهِ، وَعَقَاقِيْرَ يَسْتَعْمِلُهَا لِحِفْظِ صِحَّتِهِ،

وَبَهَائِمَ لِمَأْكَلِهِ وَلِغَيْرِ ذَلِكَ مِنْ أُمُورِهِ مِنْ حَرْثٍ وَحَمْلٍ وَغَيْرِ ذَلِكَ، وَأَزْهَارٍ وَغَيْرِهَا مِنَ العِطْرِيَّاتِ يَتَنَعَّمُ بِرَوَائِحِهَا وَيَنْتَفِعُ بِهَا، وَأَصْنَافٍ مِنَ الْمَلَابِسِ عَلَى اخْتِلَافِ أَجْنَاسِهَا، وَكُلُّ ذَلِكَ ثَمَرَةُ مَا خُلِقَ فِيْهِ مِنَ الْعَقْلِ وَالْفَهْمِ، فَانْظُرْ مَاذَا رَكَّبَ اللهُ فِيْهِ مِنَ الْعَجَائِبِ.

وَمِنَ الْحِكْمَةِ الْبَالِغَةِ اخْتِلَافُ الْعِبَادِ فِي تَمَلُّكِ مَا يَنْتَفِعُ بِهِ بَنُو

From His Transcendent Wisdom are the differences (you see) of His servants in material possessions. The children of Ādam benefit (from this difference) through the differentiation of the poor and the rich among them.[clvii] That (difference) becomes a means for the establishment of the world; by means of this, people become concerned about that which harms them in all states.[clviii] The example of what concerns them is like the example of a child, for indeed he becomes preoccupied – due to the incompleteness of his intelligence – concerning that which may harm him; (the child) is not idle, otherwise his idleness would ruin him.[clix]

How much can the one who counts enumerate from Divine Wisdoms and Subtleties (*laṭā'if*), those through which the management of the universe and its affairs have been established for a fixed term? (These wisdoms and bounties) are something that cannot be placed under any boundary or limit, and numbers cannot enumerate (their vastness). No one knows the ending of its realities and the enumeration of all truths except the Wise, the Knower – He Whose Mercy and Knowledge encompasses all things, and He who calculated all things with exquisite precision.

A CONCLUSION FOR THIS TREATISE CONCERNING THE ENNOBLEMENT OF MAN

Know that the Creator (may He be Glorified and Exalted) dignified this human being and conferred nobility upon him, as He said: "*Indeed, we have honored the Children of Ādam, and We carried them on the land and the sea, and have made provision of good things for them, and have preferred them above many of those whom We created with a marked preferment.*"[clx] So, intelligence (*'aql*) is the greatest thing through which He ennobled and honored him, as it is a faculty that allows him to recognize (and enjoy) beauty. By its

آدَمَ؛ لِيَتَمَيَّزَ مِنْهُمُ الْفَقِيْرُ مِنَ الْغَنِيِّ، فَيَكُونَ ذَلِكَ سَبَباً لِعِمَارَةِ هَذِهِ الدَّارِ، وَيَشْتَغِلَ النَّاسُ بِسَبَبِ ذَلِكَ عَمَّا يَضُرُّهُمْ فِي غَالِبِ الْأَحْوَالِ، فَمِثَالُهُمْ فِيْمَا اشْتَغَلُوا بِهِ مِثَالَ الصَّبِيِّ فَإِنَّهُ يَشْتَغِلُ لِنَقْصِ عَقْلِهِ فِيْمَا يَضُرُّ بِهِ نَفْسَهُ، وَلَا يَتَفَرَّغُ فَيَكُوْنَ فَرَاغُهُ وَبَالاً عَلَيْهِ.

وَكَمْ عَسَى أَنْ يَعُدَّ العَادُّ مِنَ الحِكَمِ وَاللَّطَائِفِ الَّتِي يُقْصَدُ بِهَا قِوَامُ الْعَالَمِ وَعِمَارَتُهُ إِلَى الْأَجَلِ الْمَعْلُوْمِ، وَهِيَ مِمَّا لَا تَدْخُلُ تَحْتَ حَدٍّ. وَلَا يَحْصُرُهَا عَدٌّ، وَلَا يَعْلَمُ مُنْتَهَى حَقَائِقِهَا وَإِحْصَاءَ جُمْلَتِهَا إِلَّا الْحَكِيْمُ الْعَلِيْمُ الَّذِي وَسِعَتْ رَحْمَتُهُ وَعِلْمُهُ كُلَّ شَيْءٍ، وَأَحْصَى كُلَّ شَيْءٍ عَدَداً.

خَاتِمَةٌ لِهَذَا الْبَابِ فِي تَكْرِيمِ الإِنْسَانِ

اِعْلَمْ أَنَّ الْبَارِيَ سُبْحَانَهُ وَتَعَالَى شَرَّفَ هَذَا الْآدَمِيَّ وَكَرَّمَهُ، فَقَالَ سُبْحَانَهُ: {وَلَقَدْ كَرَّمْنَا بَنِي آدَمَ وَحَمَلْنَاهُمْ فِي الْبَرِّ وَالْبَحْرِ وَرَزَقْنَاهُم مِّنَ الطَّيِّبَاتِ وَفَضَّلْنَاهُمْ عَلَى كَثِيرٍ مِّمَّنْ خَلَقْنَا تَفْضِيلاً} [الإسراء: ٧٠]. فَكَانَ مِنْ أَعْظَمِ مَا شَرَّفَهُ بِهِ وَكَرَّمَهُ الْعَقْلُ الَّذِي تَنَبَّهَ بِهِ عَلَى الْبَهْجَةِ، وَأَلْحَقَهُ بِسَبَبِهِ بِعَالَمِ الْمَلَائِكَةِ، حَتَّى تَأَهَّلَ بِهِ لِمَعْرِفَةِ بَارِئِهِ وَمُبْدِعِهِ

existence, (intelligence) joins him with the ranks of Angels, until he is qualified by it to (arrive at) the true recognition of his Creator and His creation by looking at His creatures. It serves as a means of proof to recognize His Qualities by way of the wisdom and trustworthiness that He placed within him.[clxi] God the Great said, "*And in your own selves, do you not see*?"[clxii] Man's contemplation of himself, and (his recognition of) that intellect that the Creator placed in him, is something that confirms his existence. Yet he is unable to fully comprehend (that)[clxiii] the intellect is one of the greatest proofs of the existence of his Maker, Planner, Creator, and Fashioner.[clxiv]

Indeed, he should reflect about his intellect and how it contains an ability to deliberate; varieties of knowledge (contained therein); a continuous recognition; keen discernments of wisdom; and an ability to distinguish between beneficial and harmful things.[clxv] Yet, despite the certainty of its existence, (marvel that he believes in it but) does not see any person or form (for it), nor does he hear any whispers (from it), nor physically perceive any tangible presence, nor smell any smell, nor come upon any form or taste (as a proof of it).[clxvi,clxvii] (The intellect) is a commander that is obeyed; it promotes and increases (in its rational abilities); it thinks and witnesses phenomena from the unseen, yet it is deluded about trivial matters. By using his intelligence, whatever is constricted from sight is broadened for him, and whatever is constricted from his cognizance has been made vast for him. He believes in whatever the barriers of God (may He be Exalted) have concealed from what is between His heavens and what exists above it, and between the earth and what exists below it, until he becomes like a witness who sees more clearly than even the eyewitness.[clxviii]

The intellect is the seat of wisdom and the storehouse of knowledge. Every time it increases his knowledge, he increases (his faculty of mental) capacity and intellectual power. (In turn) it orders

بِالنَّظَرِ فِي مَخْلُوْقَاتِهِ، وَاسْتِدْلَالِهِ عَلَى مَعْرِفَةِ صِفَاتِهِ بِمَا أَوْدَعَهُ فِي نَفْسِهِ مِنْ حِكْمَةٍ وَأَمَانَةٍ، قَالَ اللهُ الْعَظِيْمُ: {وَفِي أَنفُسِكُمْ أَفَلاَ تُبْصِرُونَ} [الذاريات: ٢١] فَكَانَ نَظَرُ الْإِنْسَانِ فِي نَفْسِهِ،

وَفِيْمَا أَوْدَعَ الْبَارِي سُبْحَانَهُ فِيْهِ مِنَ الْعَقْلِ الَّذِي يَقْطَعُ بِوُجُوْدِهِ فِيْهِ، وَيَعْجَزُ عَنْ وَصْفِهِ مِنْ أَعْظَمِ الدَّلَالَاتِ عِنْدَهُ عَلَى وُجُوْدِ بَارِئِهِ وَمُدَبِّرِهِ وَخَالِقِهِ وَمُصَوِّرِهِ،

فَإِنَّهُ يَنْظُرُ فِي الْعَقْلِ كَيْفَ فِيْهِ التَّدْبِيْرُ وَفُنُوْنُ الْعِلْمِ وَمُسْتَقَرُّ الْمَعْرِفَةِ وَبَصَائِرُ الْحِكْمَةِ وَالتَّمْيِيْزُ بَيْنَ النَّفْعِ وَالضُّرِّ، وَهُوَ مَعَ الْقَطْعِ بِوُجُوْدِهِ لَا يَرَى لَهُ شَخْصاً، وَلَا يَسْمَعُ لَهُ حِسّاً، وَلَا يَجِسُّ لَهُ مَجَسّاً، وَلَا يَشَمُّ لَهُ رِيحاً، وَلَا يُدْرِكُ لَهُ صُوْرَةً وَلَا طَعْماً، وَهُوَ مَعَ ذَلِكَ آمِرٌ وَمُطَاعٌ، وَرَاجٍ زِيَادَةً، وَمُفَكِّرٌ وَمُشَاهِدٌ لِلْغُيُوْبِ، وَمُتَوَهِّمٌ لِلْأُمُوْرِ، اِتَّسَعَ لَهُ مَا ضَاقَ عَنِ الْأَبْصَارِ.

وَوَسِعَ لَهُ مَا ضَاقَتْ عَنْهُ الْأَوْعِيَةُ، يُؤْمِنُ بِمَا غَيَّبَتْهُ حُجُبُ اللهِ سُبْحَانَهُ مِمَّا بَيْنَ سَمَاوَاتِه وَمَا فَوْقَهَا، وَأَرْضِهِ وَمَا تَحْتَهَا، حَتَّى كَأَنَّهُ يُشَاهِدُهُ أَبْيَنَ مِنْ رَأْيِ الْعَيْنِ، فَهُوَ مَوْضِعُ الْحِكْمَةِ وَمَعْدِنُ الْعِلْمِ، كُلَّمَا ازْدَادَ عِلْماً ازْدَادَ سَعَةً وَقُوَّةً، يَأْمُرُ الْجَوَارِحَ بِالتَّحَرُّكِ، فَلَا يَكَادُ أَنْ

the limbs towards action (based upon this knowledge), so much so that it is nearly incapable of distinguishing between determination and subsequent action and between subsequent action and the desire towards swiftness in obedience; this is until it becomes difficult to determine which comes first.[clxix] If determination comes first, then it is incapable of knowing itself despite its planning, knowledge, and wisdom, since it is not possible for it to describe itself with itself, with its description, or with its appearance, other than simple acknowledgements. In turn, it submits to the One who described Himself with Divine Knowledge. Man's intellect confirms this through ignorance of itself.[clxx] Now, despite his ignorance of himself, he is knowledgeable and wise, and is able to discern between the subtleties of Divine Planning. He can distinguish between the fine details of creation; how affairs run accordingly and have been managed, and how the consequences (of events) have been imagined and represented accordingly. This is clearly evident in all matters despite their variations.

Therefore, his ignorance about himself, what he knows about what is ordered, and what may be discerned, proves that he is constricted, created, fashioned, planned, and overpowered (by God). Indeed, despite his wisdom and enlightened sight, he is incapable and despicable: when he wishes to remember something, he forgets; when he wishes to forget something, he remembers.[clxxi] When he wishes to be content and happy, he grieves; when he wishes to be heedless, he is reminded. When he wishes to be aroused and awakened (to do good), he becomes inattentive and heedless. (All of these phenomena) are proofs that he is overpowered and overwhelmed. Despite what he knows, man is ignorant about the realities of what he knows. Despite what he carefully devises, he cannot understand the limit of his voice, how it exits (from him), or how there is consistency in the letters of his speech. Nor can he

يُمَيِّزَ بَيْنَ الْهَمِّ بِالْحَرَكَةِ، وَبَيْنَ التَّحَرُّكِ بِسُرْعَةِ الطَّاعَةِ أَيُّهمَا أَسْبَقُ.

وَإِنْ كَانَ الْهَمُّ قَبْلُ، وَهُوَ مَعَ تَدْبِيْرِهِ وَعِلْمِهِ وَحِكْمَتِهِ عَاجِزٌ عَنْ مَعْرِفَةِ نَفْسِهِ؛ إِذْ لَا يُمْكِنُهُ أَنْ يَصِفَ نَفْسَهُ بِنَفْسِهِ بِصِفَةٍ وَهَيْئَةٍ أَكْثَرَ مِنَ الْإِقْرَارِ بِأَنَّهُ مُسَلِّمٌ لِلَّذِي وَصَفَهُ العَلِيْمُ بِهِ، وَمُقِرٌّ بِالْجَهْلِ بِنَفْسِهِ وَهُوَ مَعَ جَهْلِهِ بِنَفْسِهِ عَالِمٌ حَكِيْمٌ، يُمَيِّزُ بَيْنَ لَطَائِفِ التَّدْبِيْرِ، وَيُفَرِّقُ بَيْنَ دَقَائِقِ الصُّنْعِ، وَتَجْرِيْ الْأُمُوْرُ وَقَدْ تَدَبَّرَهَا، وَيَتَوَهَّمُ العَوَاقِبَ وَقَدْ تَمَثَّلَهَا، وَيَدُلُّ عَلَى الْأُمُوْرِ عَلَى اخْتِلَافِهَا، فَدَلَّ جَهْلُهُ بِنَفْسِهِ وَعِلْمُهُ بِمَا يُدَبِّرُ وَيُمَيِّزُ أَنَّهُ مُرَكَّبٌ مَصْنُوْعٌ مُصَوَّرٌ مَقْهُورٌ؛

لِأَنَّهُ مَعَ حِكْمَتِهِ وَاتِّقَادِ بَصِيْرَتِهِ عَاجِزٌ مَهِيْنٌ، يُرِيْدُ أَنْ يَذْكُرَ الشَّيْءَ فَيَنْسَاهُ، وَيُرِيْدُ أَنْ يَنْسَاهُ فَيَذْكُرَهُ، وَيُرِيْدُ أَنْ يُسَرَّ فَيَحْزَنَ، وَيُرِيْدُ أَنْ يَغْفُلَ فَيَذْكُرَ. وَيُرِيْدُ أَنْ يَنْتَبِهَ وَيَتَيَقَّظَ فَيَسْهُوَ وَيَغْفُلَ؛

دَلَالَةً عَلَى أَنَّهُ مَغْلُوْبٌ مَقْهُوْرٌ، جَاهِلٌ بِحَقَائِقِ مَا عَلِمَ، وَهُوَ مَعَ مَا دَبَّرَ لَا يَدْرِيْ كَمْ مَدَى مَبْلَغِ صَوْتِهِ، وَلَا كَيْفَ خُرُوجُهُ، وَلَا كَيْفَ اتِّسَاقُ حُرُوْفِ كَلَامِهِ،

وَلَا كَمْ مَدَى مَبْلَغِ نَظَرِهِ، وَلَا كَيْفَ رُكِّبَ نُوْرُهُ، وَلَا كَيْفَ أَدْرَكَ الْأَشْخَاصَ، وَلَا كَمْ قَدْرُ قُوَّتِهِ، وَلَا كَيْفَ تَرَكَّبَتْ إِرَادَتُهُ وَهِمَّتُهُ؟

understand the extent of his sight, the construction of light, or the process by which he comes to perceive shapes and figures (through vision). Nor does he know the limit of his strength or how both intention and motivation (towards accomplishing a given task) come together in him. Therefore, using his knowledge, he attempts to prove the reality of what he knows *a priori* that he has been created in a fine state and with a permeating (Divine) Wisdom by the Maker, the Creator, the Desirer, the Knower (may He be Exalted and Glorified).

The faculty of desire has been created in man such that it is congruent with his nature. If he seeks to employ the light of his intelligence towards purposes for which he was commanded, he would arrive safely and triumph in the house of nobility tomorrow (i.e., the Hereafter).[clxxii] However, if he seeks to use his desire towards the purposes of his lower self and its desires, he will be veiled from the true recognition of affairs. He will never comprehend this despite having a powerful intellect, and he will be barred from this recognition in the Hereafter by instead receiving his due recompense, veiling, and punishment.[clxxiii]

Thus, it is a tool (for him) in industrial work and in daily tasks (as well), such that he measures and plans in his mind and imagination, and then extrapolates all that he has imagined, deducing by way of fine and subtle thinking and by recognizing the virtuous characteristics that are present in every community and time.[clxxiv] (Through intellect) he is able to deem beautiful and excellent the traditions of intellectuals and noblemen that are deserving (of being appreciated), and he is able to deem loathsome what is deserving (to be labeled) among them through his discernment of (their) customs.[clxxv]

Then ponder upon how God honored this human being and created faculties in him that benefit him towards (learning and

فَاسْتَدَلَّ بِعِلْمِه - عَنْ حَقِيْقَةِ مَا عَلِمَ - أَنَّهُ مَصْنُوْعٌ بِصَنْعَةٍ مُتْقَنَةٍ، وَحِكْمَةٍ بَالِغَةٍ تَدُلُّ عَلَى الصَّانِعِ الْخَالِقِ، الْمُرِيْدِ الْعَلِيْمِ عَزَّ وَجَلَّ.

ثُمَّ إِنَّهُ خَلَقَ فِي الْإِنْسَانِ الْهَوَى مُوَافِقاً لِطِبَاعِهِ، فَإِنِ اسْتَعْمَلَ نُوْرَ الْعَقْلِ فِيْمَا أُمِرَ بِهِ وَرَدَ مَوْرِدَ السَّلَامَةِ، وَفَازَ غَداً بِدَارِ الْكَرَامَةِ، وَإِنْ اسْتَعْمَلَهُ فِي أَغْرَاضِ نَفْسِهِ وَهَوَاهَا حُجِبَ عَنْ مَعْرِفَةِ أُمُوْرٍ لَا يُدْرِكُهَا غَيْرُهُ، مَعَ مَا هُوَ مُتَوَقَّعٌ لَهُ فِي الدَّارِ الْآخِرَةِ مِنَ الثَّوَابِ وَالْحِجَابِ وَالْعِقَابِ.

وَهُوَ الْآلَةُ فِي عَمَلِ الصَّنَائِعِ وَتَقْدِيْرِهَا عَلَى نَحْوِ مَا قَدَّرَهَا وَدَبَّرَهَا فِي ذِهْنِهِ وَتَخَيُّلِهِ، وَاسْتِنْبَاطِ مَا يُسْتَنْبَطُ بِدَقِيْقِ الْفِكْرِ وَمَعْرِفَةِ مَكَارِمِ الْأَخْلَاقِ الْمَوْجُوْدَةِ فِي كُلِّ أُمَّةٍ وَزَمَانٍ، وَاسْتِحْسَانِ مَا يَحْسُنُ فِي عَوَائِدِ الْعُقَلَاءِ وَالْفُضَلَاءِ، وَتَقْبِيْحِ مَا يَقْبُحُ عِنْدَهُمْ بِحُكْمِ الِاعْتِيَادِ.

فَانْظُرْ مَا شَرَّفَ اللهُ بِهِ هَذَا الْإِنْسَانَ، أَنْ خَلَقَ فِيْهِ مَا يُفِيْدُهُ هَذِهِ الْمَعَارِفَ؛ فَإِنَّ الْأَوَانِيَ تَشْرُفُ بِشَرَفِ مَا يُوْضَعُ فِيْهَا. وَلَمَّا كَانَتْ قُلُوْبُ الْعِبَادِ هِيَ مَحَلٌّ لِلْمَعْرِفَةِ بِاللهِ سُبْحَانَهُ وَتَعَالَى شَرُفَتْ بِذَلِكَ،

وَلَمَّا سَبَقَ فِي عِلْمِ الْبَارِيْ سُبْحَانَهُ وَإِرَادَتِهِ وَحِكْمَتِهِ بِمَصِيْرِ الْخَلْقِ إِلَى دَارٍ غَيْرِ هَذِهِ الدَّارِ. وَلَمْ يَجْعَلْ فِي قُوَّةِ عُقُوْلِهِمْ مَا يَطَّلِعُوْنَ بِهِ عَلَى

attaining) these recognized virtues. Indeed, vessels are honored according to what is placed in them. Since the hearts of the servants are the place for recognizing God ﷻ, they have been honored with that (ability).

Since it came to pass in the Knowledge, Will, and Wisdom of the Creator ﷻ that the destiny of creation is towards an abode other than this abode, and since He did not place in their intellectual potential any ability to stumble upon the laws of that world, He instead perfected this light of intellect – which He bestowed on them – with the light of Prophethood (as a further favor). Thus, He sent them Messengers ﷺ who came to bring glad tidings for the people of obedience and warn the people of disobedience. He provided them with revelation and prepared them to accept and receive it. Therefore, the lights of what came in revelation from God ﷻ were joined to the light of intellect, just as the sun is joined to the light of a star.[clxxvi] They guided mankind towards the best of their worldly affairs that the human mind cannot independently derive, and directed them towards the best of their otherworldly affairs – a realm about which there is no recourse for mankind to understand except through the medium (of Prophets ﷺ). He made apparent to them the veracity of the proofs that Prophets ﷺ came with – proofs that necessitated submission and obedience due to the veracity of their statements. Through them, the favour of God was complete upon His servants, His Generosity was made apparent, and His Proof against them (if they disbelieved) was definitively established.

Now observe what (favors) man and his progeny have been dignified with, such that from them came noblemen who became recipients of these meritorious additions. Then see how uniquely different Divine Laws came to humanity (through known Prophets), like different types of light that come from the Sun.[clxxvii] The lights of intellects also differed like the lights of the stars (yet they all

أَحْكَامِ تِلْكَ الدَّارِ، كَمَّلَ لَهُمْ سُبْحَانَه هَذَا النُّورَ الَّذِي وَهَبَهُمْ إِيَّاهُ بِنُوْرِ الرِّسَالَةِ إِلَيْهِمْ، فَأَرْسَلَ الْأَنْبِيَاءَ صَلَوَاتُ اللهِ عَلَيْهِمْ مُبَشِّرِيْنَ لِأَهْلِ طَاعَتِهِ، وَمُنْذِرِيْنَ لِأَهْلِ مَعْصِيَتِهِ، فَمَدَّهُمْ بِالْوَحْيِ،

وَهَيَّأَهُمْ لِقَبُوْلِهِ وَتَلَقِّيْهِ، فَكَانَتْ أَنْوَارُ مَا جَاءَ بِالْوَحْيِ مِنْ عِنْدِ اللهِ بِالنِّسْبَةِ إِلَى نُوْرِ الْعَقْلِ كَالشَّمْسِ بِالْإِضَافَةِ إِلَى نُوْرِ النَّجْمِ، فَدَلُّوا الْعِبَادَ عَلَى مَصَالِحِ دُنْيَاهُمْ فِيمَا لَا تَسْتَقِلُّ بِإِدْرَاكِهِ عُقُوْلُهُمْ،

وَأَرْشَدُوْهُمْ إِلَى مَصَالِحِ أُخْرَاهُمُ الَّتِي لَا سَبِيْلَ لِلْعِبَادِ أَنْ يَعْرِفُوْهَا إِلَّا بِوَاسِطَتِهِمْ، وَأَظْهَرَ لَهُمْ سُبْحَانَهُ مِنَ الدَّلَائِلِ عَلَى صِدْقِ مَا جَاؤُوا بِهِ مَا أَوْجَبَ الْإِذْعَانَ وَالِانْقِيَادَ لِصِدْقِ أَخْبَارِهِمْ، فَتَمَّتْ بِذَلِكَ نِعْمَةُ اللهِ عَلَى عِبَادِهِ، وَظَهَرَتْ كَرَامَتُهُ وَثَبَتَتْ حُجَّتُهُ عَلَيْهِم.

فَانْظُرْ مَا أَشْرَفَ الْآدَمِيَّ وَنَسْلَهُ الَّذِيْنَ ظَهَرَتْ مِنْهُمْ هَؤُلَاءِ الْفُضَلَاءُ الَّذِيْنَ هُمْ قَابِلُوْنَ لِهَذِهِ الزِّيَادَاتِ الْفَاضِلَةِ، ثُمَّ تَضَافَرَتْ أَنْوَارُ الشَّرَائِعِ الَّتِيْ هِيَ كَالشَّمْسِ، وَأَنْوَارُ الْعُقُوْلِ الَّتِي هِيَ كَالنَّجْمِ، فَتَمَّتْ سَعَادَةُ مَنْ سَبَقَ لَهُ مِنَ اللهِ الحُسْنَى، وَشَقَاوَةُ مَنْ كَذَّبَ وَلَمْ يُرِدْ إِلَّا الْحَيَاةَ الدُّنْيَا.

ثُمَّ إِنَّ اللهَ تَبَارَكَ وَتَعَالَى مَنَّ عَلَى الْإِنْسَانِ بِأَنْ خَصَّهُ بِرُؤْيَا يَرَاهَا

reached the same conclusion about the Oneness of God and man's accountability in the Hereafter). How complete (and perfect) is the happiness of one for whom God has destined goodness, and (complete) is the wretchedness of one who denies this and desires only the life of this world!

Then God (blessed is He and Exalted) further favoured man by exclusively giving him the ability to dream and see in his sleep – or even to see (while awake) with his eyes like a dream – forms and figures that were previously recognized (while awake).[clxxviii] These (true) dreams are either glad tidings or warnings, according to what occurred to him.[clxxix] All of these are noble gifts and favors from the generosity of God (may He be Glorified). God then made his steadfastness upon obedience in his heart and limbs to be a cause for the veracity of these gifts, permeating all things. This is so that man may take heed and plan ahead his affairs or save himself from dangers. These are affairs about which God has exclusively reserved knowledge (to Himself). Sometimes, He may share some of this knowledge with those whom He wishes.

فِي مَنَامِهِ أَوْ فِي عَيْنِهِ كَشِبْهِ الْمَنَامِ يُمَثِّلُ لَهُ فِيْهَا بِأَمْثِلَةٍ مَعْهُوْدَةٍ مِنْ جِنْسِ مَا يَعْرِفُهُ. وَهِيَ مُبَشِّرَةٌ أَوْ مُنْذِرَةٌ لَهُ لِمَا يَتَوَقَّعُهُ بَيْنَ يَدَيْهِ، كُلُّ ذَلِكَ مَوَاهِبُ وَكَرَامَاتٌ مِنْ جُوْدِ اللهِ سُبْحَانَهُ.

وَجَعَلَ اللهُ اسْتِقَامَتَهُ عَلَى الطَّاعَةِ فِي قَلْبِهِ وَجَوَارِحِهِ سَبَباً لِصِدْقِهَا فِي غَالِبِ الْأَمْرِ؛ لِيَتَّعِظَ أَوْ يُقْدِمَ عَلَى الْأُمُوْرِ أَوْ يُحْجِمَ عَنْهَا، وَهِيَ الْأُمُوْرُ الَّتِي انْفَرَدَ اللهُ بِعِلْمِ الْعَاقِبَةِ فِيْهَا، وَأَطْلَعَ عَلَى بَعْضِ الْأُمُوْرِ مِنْهَا مَنْ شَاءَ مِنْ عِبَادِهِ.

الحمد لله، الذي أتم هذا كتابي
وما فعلته عن أمري
اللهم تقبّل مني، إنك أنت السميع العليم.
الفقير إلى الله: كامران محمد رياض

O Lord, this task of mine is now complete,
With mistakes and errors, I pray, it is not replete.
May it be a light for all those who read,
And benefit me when I am in need.
- KMR

Endnotes

i **Publisher's Note:** Clarifications of the text are in unitalicized brackets. The initials MAK (for Sheikh Mohammed Amin Kholwadia) and KMR (for Kamran M. Riaz) are placed after notes to distinguish between the comments. Notes without an initial denote references or clarifications of the translation.

ii "Indeed We created man from an extract of clay, then we placed him as a (drop of) sperm in a safe lodging" (*al-Mu'minūn*, 12-13).

iii It is worth noting that the author uses some elements of passive voice in the opening paragraphs; not the majhūl, but rather an indication that the event of the "Creation of the created" came to pass. Nonetheless, God ﷻ is the Agent of all creation, placing them on the Earth, etc. Thus, the author is subtly indicating the Divine's role in this process. But it is also understood that the zygote itself is not "alive", as the process of ensoulment does not occur until approximately 120 days after fertilization, as understood from a famous Prophetic statement (narrated from ʿAbdullāh ibn Masʿūd ؓ by al-Bukhārī ؒ, *Ṣaḥīḥ al-Bukhārī*, hadith #3208, and Muslim ؒ, *Ṣaḥīḥ Muslim*, hadith #2643). It is biologically alive, but it is more appropriate, under a metaphysical framework, to understand that at this point it contains the potential for complete human life. (MAK)

iv Sperm cells contain genetic material capable of producing every cell in the body. Thus, one cell contains information that is the essence of the human body. A single sperm cell, which is the result of a process of cell division called spermatogenesis, contains the paternal half of the genetic material (23 chromosomes) that comprises human somatic cells. A similar cell division termed oogenesis occurs in females, which produces oocytes that contain the maternal contribution (23 chromosomes) to somatic cells. Sperm cells are capable of motility via flagella that allow them to swim towards the egg for fertilization using specific movements while traversing a hostile environment. The structure, mobility, and biophysical properties of the sperm cell – such as possessing an acrosome to neutralize egg cell defences and permit fertilization – allow the sperm to reach and fertilize the egg. The sperm and egg fuse to form a zygote with a complete set of 46 chromosomes that comprise the human genome. Thus, the zygote contains sufficient genetic coding for the development and functioning of the human body [Moore and Persaud: pp. 16-20, 31-35]. (KMR)

v The word used here is bi sababi al-iflāj. The root verb, fa-la-ja, means "to split, to cleave" as an active verb, but it can also mean "to be semi-paralyzed" if read as a passive verb. Biologically, the zygote is actively changing but also undergoes passive movement from one location to another. Another reading of this word may be īlāj, which translates to "insertion", meaning that the zygote is carried through the process of insertion from the male to the female (one hidden location to another).

vi This refers to movement from the site of implantation. Fertilization normally occurs in the upper third of the fallopian tube as the ovum is descending from the ovary towards the uterus. The zygote then travels in a retrograde fashion to cling to the wall of the uterus.

vii Despite the hostile environment it faces and the circuitous anterograde and retrograde motion that the zygote must undergo, it remains capable enough to develop life, unaffected by the climate around it. Al-Ghazālī ﷺ is inviting the reader, especially the one who is familiar with these biological processes, to reflect on a seemingly commonplace yet extraordinary process that starts in the father's body and harmoniously culminates in the mother's body. (MAK)

viii The author implies here that any dissimilarity would therefore not favor the development of human life.

ix "Then We fashioned the sperm into a clot of congealed blood; then of that clot, We made a lump of flesh; then We made bones in that lump; then We clothed the bones with flesh; then We developed out of it another creature. So blessed be God, the best of creators" (al-Mu'minūn, 14). (MAK)

x The author astutely notes both a physiological and metaphysical phenomenon. Physiologically, the eye is an organ that significantly depends on the extent to which it is used. It may adjust to the darkest and brightest of environments, accommodate the nearest and farthest of objects, and comprehend the entire spectrum of visible colours. Its very health also depends on a variety of factors, including how often it is used. For example, vision may not develop in a child if the eye cannot see well (amblyopia), or it may atrophy when it is not used (deprivational vision loss). These conditions often require a significant period of rehabilitation to regain former function, and in some cases, may be irreversible. (KMR)

xi There are three points worthy of mention here. Metaphysically, the author uses the eye's potential for mystical vision, both in ru'yah (Divinely-inspired inner vision) and in dreams – topics that he will address at the culmination of this treatise – to indicate the eye's potential ability to allow the human being to witness metaphysical sights as proofs of his journey to become closer to God ﷻ. He is perhaps indicating towards the concept of qurrat aʿyun (the felicity of the eye), wherein the eye – as well as the human being – derives pleasure from witnessing physically and metaphysically beautiful sights.

Secondly, he subtly introduces the idea that the eye itself needs assistance in order to reveal things. He introduces this theme here, though he will revisit it later in the treatise when he marvels at the various mediums created by God ﷻ that allow the senses to be employed properly, and how their absence would render human senses useless. Finally, the consciously abstruse language employed by the author suggests that the spiritual benefits of the eye cannot be elucidated; he invites the reader to instead reflect, as this is the first step in experiencing these benefits. (MAK)

xii Our current understanding of the anatomical layers of the eye (excluding the eyelids) from outside to inside are roughly: the conjunctiva, fascia bulbi (Tenon's capsule), sclera, choroid, and retina. Anteriorly, the (seven) layers of the eyelid are: skin, orbicularis oculi muscle, orbital septum, preaponeurotic fat, levator aponeurosis, Muller's muscle, and tarsal conjunctiva. It is unclear if the author refers to these seven layers as in the layers of the Earth (e.g., crust, mantle, outer core, inner core) or anteriorly-posteriorly in an axial direction. If the latter, a simplified seven layers breakdown would be: eyelid, cornea, iris, lens, vitreous, retina, and optic nerve. (KMR)

xiii For example, excessive thickness in a layer, such as corneal edema (swelling), cataract lens (as occurs with aging), and retinal edema (as may occur in systemic conditions such as diabetes, or prolonged ocular inflammation) can severely hinder vision. (KMR)

xiv Both trichiasis (abnormal growth of eyelashes that grow inwards) and distichiasis (an extra row of eyelashes) can irritate the ocular surface. Madarosis (loss of eyelashes) can occur due to both ocular pathology and systemic conditions. Madarosis may sometimes be the first (and only) clinical sign of systemic diseases. (KMR)

xv Tear fluid contains high amounts of albumin and immunoglobulins (predominantly IgA, IgG, and IgE). The average pH level is 7.35 (range 5.20-8.35), and usually slightly basic. While the tear film has several functions, its two main functions are to inhibit the growth of microorganisms by mechanical flushing and antimicrobial action, as well as to moisture and protect the delicate surface of the corneal and conjunctival epithelium. (KMR)

xvi There is a spherical curvature to the part of the eye that faces the outside world. Anything that falls into the eye is directed towards either the medial or lateral canthus, and is therefore less likely to damage the cornea, which is in the center of the eye. Even a small wound or insult to the cornea would obstruct vision or harm the eye. (KMR)

xvii This is a fascinating observation. Eyebrow and eyelash hair grow to a certain length and then stop growing. They do not require any grooming or intervention from the human being to continue its aforementioned functions, whereas hair on the scalp and beard require active maintenance. (KMR)

xviii It is worth mentioning here that the word "then" (thumma) is used multiple times in the course of the author's anatomical survey. This is used more for the sense of going "down" the body systematically rather than implying a chronological order of the creation of body parts or a hierarchy of human anatomy.

xix The author used the analogy of a door in the context of the eye earlier, and will use this in reference to the lips later. The significance of this is unclear, and it may very well be a coincidence. (MAK)

xx While he states that teeth are "not (like) a single bone", it may be inferred that there are both similarities and differences between teeth and bones. For example, teeth resemble bones in sturdiness and composition, as teeth possess large concentrations of calcium. Yet the consistency of teeth allows them to undergo repetitive mechanical motion with forces that are not experienced by bones [see Table 1]. While teeth are more mineralized to provide crush resistance from repetitive forceful maneuvers of the jaw, bones have a more flexible element that allows them to remain hard yet be lightweight to perform many other important functions such as support, movement, mineral storage, and hematopoiesis [Kierszenbaum: pp. 147-48, 433-37; Moore: pp. 382, 494]. (KMR)

xxi Saliva is composed primarily of water and electrolytes, but also contains important substances, including but not limited to: antibacterial compounds such as IgA (a secretory immunoglobulin) and hydrogen peroxide; digestive enzymes (such as amylase and lipase) and other minor enzymes (such as lingual lipase, lysozyme, and kallikrein); other substances such as opiorphin, a natural pain killer [Guyton: pp. 740-41]. (KMR)

xxii Canines and incisors are involved in cutting food, thereby allowing other teeth such as molars to perform tasks such as crushing and grinding food to render it suitable for swallowing. Canines have long roots and sharp crowns that facilitate the tearing of food. Molar teeth have more roots that are attached to the alveolar bone, making them stronger for crushing food [Moore & Agur: p. 555-56, Anderson: plates 7-97-101]. (KMR)

xxiii This is a quintessential example in the treatise of the author's use of the scientific method. He makes a two-step observation: the first is that saliva exists, and a second observation indicating under what conditions it is produced. This is quite intuitive, as saliva release is uniquely regulated and controlled by both parasympathetic and sympathetic stimulation. Sympathetic stimulation preferentially produces thicker saliva, whereas parasympathetic stimulation produces thinner saliva. Each of these two types of saliva has different percentages of water and electrolyte content. In addition, the rate of saliva flow also determines the composition of saliva. The salivary gland consists of acini and ducts, both of which produce saliva with a unique composition. For example, saliva secreted from the acini is mostly isotonic with an electrolyte concentration similar to plasma, which then drains through intercalated/excretory ducts that modify this electrolyte composition, extracting sodium (Na) and chloride (Cl) and adding potassium (K) and bicarbonate. Therefore, at low saliva flow rates, saliva is hypotonic, slightly acidic, and rich in K, but depleted in Na and Cl (as the ducts have more time to modify the saliva). At higher flow rates, saliva is isotonic and contains electrolyte concentrations similar to those found in plasma [Ganong: Chapter 26]. (KMR)

xxiv The author astutely notes this positive feedback system in the human body, because saliva is released in the presence of food, which in turn stimulates the production of even more saliva. Physiologically, saliva consists of many substances as previously mentioned. An example of this is kallikrein, which aids in the conversion of high-molecular-weight kininogen into bradykinin, a potent vasodilator that increases blood flow to salivary glands, thereby producing more saliva [Costanzo: p. 317]. (KMR)

xxv This correlates well with a disease known as Sjögren's syndrome, an autoimmune condition where immune cells produce auto-antibodies (specifically, antibodies to Ro(SSA) or La(SSB) antigens, or both) that destroy the exocrine glands that produce tears and saliva. Patients primarily exhibit a dry mouth and dry eyes (sicca symptoms), and have extreme difficulty with speech and chewing food. Dry foods may even cause choking in these patients if not taken with sufficient water. There is also an increased risk of B-cell lymphoma in these patients [Kumar: pp. 235-37]. (KMR)

xxvi How the author comes to this information is a topic worthy of discussion. He may have gained this knowledge from the leading academics of his time, from his own intuition, or through Divinely-inspired knowledge (ilhām). He perhaps provides an insight later in the treatise when he discusses the stages of epistemology, placing objective observation as a starting point, and Divinely unveiled knowledge (*kashf*) as the inspiration and pinnacle. (MAK)

xxvii The author may have also gained this knowledge through empirical observation. The following treatise in this collection of essays – Concerning Divine Wisdom in the Creation of Birds – provides further details regarding his highly sophisticated scientific method. Principles such as direct observation to support theory form a key component of al-Ghazālī's ﵀ paradigm of scientific inquiry that is inspired by faith and spiritual curiosity. One may note that his method is remarkably similar to the scientific method of René Descartes (d. 1650), and certainly predates it by several centuries. (KMR)

xxviii Likely a reference to *al-Ra ʿd*, 11.

xxix Here, the author is inviting the reader to ponder upon the taste in the food that one eats and how it is a mercy. For God ﷻ could have made food to have no taste or a foul taste, yet man would still have to eat in order to survive. He marvels that God ﷻ favored man by allowing him to enjoy a function that is required for survival. God ﷻ

says, "We preferred some (fruits) in taste above others; most surely there are signs in this for a people who understand" (*al-Ra ʿd*, 4), and the author continues this sentiment in his own observation. (MAK)

xxx Physiologically, taste pathways are amongst the most complex in the body. Animals, including humans, form particularly strong aversions to novel foods if illness follows their consumption. The survival value of such aversions is apparent in terms of avoiding poisons [Ganong: Chapter 10]. (KMR)

xxxi He marvels at how both faculties – taste and sensory perception – are united in the same organ. The tongue, in addition to containing specialized pathways for taste conduction, also possesses pathways for transmission of sense, pain and temperature; these are similar to pathways found in other parts of the body. Taste pathways are conducted via the chorda tympani of cranial nerve (CN) VII from the anterior tongue, and CN IX from the posterior tongue, with minor contribution from CN X. These unite and follow a unique tract to reach the brainstem via the nucleus tractus solitarius. Sensory innervation is slightly different in that sense perception from the anterior tongue is conducted via the CN V3 branch of the lingual nerve, and from the posterior tongue it is also conducted by CN IX. These pathways unite and are conducted via regular sensory pathways, which are different from taste pathways [Moore and Dailey: pp. 1003-5; see also FIGURE 1]. (KMR)

xxxii This is perhaps one of the most incredible intuitions of al-Ghazālī ﷺ in the entire treatise. Embryologically, the inner ear develops early in the fourth week, with the appearance of the otic placode, a surface ectoderm thickening, which appears on each side of the caudal midbrain. Each otic placode soon invaginates and migrates deep into the underlying mesenchyme, forming an otic pit. The edges of the otic pit fuse together and form an otic vesicle, which is the beginning of the membranous labyrinth. The verb he uses, shaqqa, literally means "to split, to furrow, to cleave", which is an uncannily precise description of this microscopic embryologic process [Moore and Persaud: p.472; see also FIGURE 2]. (KMR)

xxxiii He notes the usefulness of cerumen that is found in the human ear. While he is likely referring to macroscopic invaders such as parasites and insects, cerumen provides defence against even microscopic entities such as bacteria, fungi, and viruses. It is noteworthy that scientific literature as recently as 1956 found little evidence to support the antibacterial properties of cerumen [Perry: pp. 165–70]. However, there have been studies that demonstrate the bactericidal properties of cerumen, especially against bacteria that are notorious for causing otitis media, such as Haemophilus influenzae [Chai: pp. 638-641]. This is likely due to the presence of cerumen's slight acidic pH (6.1) and the presence of bactericidal enzymes such as lysozymes [Jahn: pp. 34-39]. (KMR)

xxxiv His use of the word murratan can be translated as either "bitter" or "sour". (MAK) xxxv The sensory innervation of the ear is highly dense and complex, making it one of the most sensitive areas in the human body, for the purposes of warning and protecting the human being [Levy: pp. 154, 158-67]. (KMR)

xxxvi He is saying that the wisdom in the ear having this curvature is not only to conduct the proper amount of noise from the outside world, but also so that if some insect fell into the ear, the path to reach the brain is lengthened by these curves despite the small space. The insect's movement would therefore rouse the person even from sleep so that he could protect himself from this danger. This would not be possible if the ear canal was short or straight, or if this heightened sensitivity was absent. Pathogens would have an easier access to infect the brain and central nervous system. It is also noteworthy that there is curvature of the inner ear and bony labyrinth behind the tympanic membrane. Thus, there is inherent structural and biochemical defences against inner and outer ear infections (otitis media and otitis externa, respectively). (KMR)

xxxvii It is noteworthy that the author does not comment on nostril hair or its usefulness, whereas he comments on hair in other places as seen thus far. (KMR)

xxxviii Modern research has shed light that smell not only provides us information about our environment, but is also linked to memory, emotion, and consciousness. There is speculation on why we remember or associate particular events or places with specific smells. The literature discusses why certain smells can cause the full spectrum of emotions, including depression, joy, and anger. Studies have also implemented the use of smelling salts to rouse an unconscious patient. New discoveries include an understanding of the existence and hypothesized role of pheromones on a multitude of human behaviors, especially intimacy, which make use of the olfactory system. (KMR)

xxxix There are many spiritual connections possible here. For example, and especially in light of the above footnote, there is a Prophetic hadith interpreted by the famous Tābiʿī Uways al-Qarnī ﵁ that states, "Human souls are like a host of troops: they smell (one another) like horses smell (one another)."

قَالَ : عَرَفَتْ رُوحِي رُوحَكَ ؛ لِأَنَّ أَرْوَاحَ الْمُؤْمِنِينَ تَشَامُّ كَمَا تَشَامُّ الْخَيْلُ ، فَمَا تَعَارَفَ مِنْهَا ائْتَلَفَ وَمَا تَنَاكَرَ مِنْهَا اخْتَلَفَ

(al-Asfahānī ﵀, *Ḥilyat al-Awliyā'*, 10:16 Dār al-Fikr, 2019; al-Bukhārī ﵀, *Ṣaḥīḥ al-Bukhārī*, hadith #3336, Muslim ﵀, *Ṣaḥīḥ Muslim*, hadith #2638). This suggests that there are olfactory mechanisms at work in determining compatibility and other relationships between human beings. It is also apparent that smells are instrumental even in man's relationship with Divinely created beings like Angels; several Prophetic statements encourage Muslims towards meticulous hygiene, especially during ritual prayer, as these beings of Divine Light are attracted to pleasant smells. It is perhaps no coincidence that the Prophet ﷺ said that perfume was among the three things that were made beloved to him (Narrated from Anas ibn Mālik by Imam Aḥmad ؓ, *Musnad Aḥmad*, hadith #14069 and al-Nasā'ī ؓ, *Sunan al-Nasā'ī*, hadith #3940). Finally, the epitome of the significance of this sense may be best understood through a Prophetic hadith: "*By Him in Whose Hands is my soul, the smell emanating from the mouth of a fasting person is better in the sight of God ﷻ than the smell of musk*" (al-Bukhārī, *Ṣaḥīḥ al-Bukhārī*, hadith # 7492, The Book of Fasting, hadith #118). While it is erroneous to suppose that human beings share sense-faculties with God ﷻ, one may marvel that the "closest" faculty, according to mystic-scholars such as Ibn ʿArabī (d. 1240), is smell, since God ﷻ is also pleased with beautiful smells. (MAK)

xl This is also a reference to the Prophetic statement mentioned above (on perfume) that the sense of smell, just as the sense of taste, not only provides information about man's environment, but also gives man pleasure when he encounters pleasant smells. Thus, the mercy of God ﷻ is so vast that smelling, just like eating, is not only utilitarian, but also a source of enjoyment and comfort for man. The power of smell to comfort and give joy is further epitomized in other hadith, such as one that states that a believer will smell Paradise before he enters it, and that this smell is powerful enough that he will forget all the sorrow and hardships he faced in his earthly life (al-Bukhārī, *Ṣaḥīḥ al-Bukhārī*, hadith # 6914). (MAK)

xli This is simultaneously poetic and ironic, as the "scent of life", oxygen, has no inherent smell. (KMR)

xlii The word ṣawt may also be translated as "sound" in a more literal sense. However, his point here is to stress that no two voices are similar. In a more general sense, one may also infer that no two sounds are similar as well. (MAK)

xliii "Among His Signs is the creation of the heavens and the earth, and the diversity of your tongues and colors; indeed in that are Signs for the learned" (*al-Rūm*, 22). While this verse primarily refers to the diversity of languages, it may also be inferred that the diversity of vocal sounds is a quintessential component in the development of languages.

This may explain why certain phonetic sounds are nearly universal in all languages, while other sounds are unique to a given culture or period of history. Regardless, appreciation and reflection are due in recognizing that this diversity of languages – that makes human beings unique when compared to all other creation – is also a Divine Mercy. (MAK)

xliv Here, al-Ghazālī ؓ is indicating that there was either a difference in the faces of Ādam and Eve – males and females – or in their progeny. When they were banished to the Earth, this difference allowed them to recognize one another, and also allowed their progeny to recognize male and female, self and non-self, in order to co-exist on Earth. It also made them appreciate that different faces were like other differences, such as skin colour, cultures, and tribes. These differences were all established for the purpose of recognizing one another, an allusion to the famous verse: "O mankind, indeed We made you into tribes and nations in order that you may recognize one another" (*al-Ḥujurāt*, 13). (MAK)

xlv It is mutual recognition of each other on a mundane level. But it is also meant for human beings to recognize the profound power and diversity of the One who created such things. (MAK)

xlvi Linguistically, the word yad includes the entire upper extremity from the fingertip to the armpit, but it may also be used for the hand only.

xlvii The word anāmil here refers to the individual bones (metacarpals), or parts of the finger, though it may refer to fingertips as well. I have chosen to leave the translation as finger bones in this sentence, whereas I have translated it as "fingertips" later in another sentence.

xlviii The existence of, and differences in, human fingertips are considered to be one of the most profound manifestations of the Majesty and Ability of God ﷻ: "Does man think We shall not gather his bones? Nay, We are able to put together in perfect order the very tips of his fingers!" (*al-Qiyāmah*, 3-4). (MAK)

xlix On a biological level, the existence of the opposable thumb – which allows for the performance of complex tasks requiring fine movements such as grasping – is a unique feature of human beings. His use of the word "hover" (yadūru) is also interesting, as the thumb is the only 'finger' that can circularly rotate and contact the other fingers. (KMR)

l On a spiritual level, the author mentions "hovering" and "rotating" of the thumb – this is a distinct, Prophetically taught movement for the believer to keep count of his remembrance of God ﷻ. Whereas the Qur'an speaks of all creatures remembering and praising God ﷻ in their own ways (*al-Isrā'*, 44), the physical act of using the fingers while remembering God ﷻ (*tasbīḥ*) is unique to the human being, and al-Ghazālī 🙢 marvels that the human being has the innate anatomical machinery to accomplish this. It is also worth mentioning that the Prophetically taught method of counting, particularly the procedure to keep track of tens, mainly requires the use of the thumb. (MAK)

li The author is asserting that there is no other "possible" useful arrangement of the thumb and four fingers other than what He created. (MAK)

lii The diversity and precision of the Arabic language is worth mentioning here, as there are subtle yet remarkable differences in how the author juxtaposes these nouns of instrumentation (ism ālah). The word mighfarah (ladle) means to hold one's hand with fingers together facing up, such as in a position suitable for holding water. The word mijrafah (scoop) is a similar position, with the notable exception that one's fingers are facing downwards, such that one may dig in the earth with bare hands. (MAK) liii In addition to worldly functions, the "ladle" and "scoop" hand configurations and maneuvers allow the believer to make ablution (*wuḍū'*) and supplication (*du ʿā'*), respectively. (MAK)

liv It was a Prophetic habit and a meritorious act for the male believer to keep his nails closely clipped (narrated by Abū Ḥurayrah in *Ṣaḥīḥ al-Bukhārī*, hadith #5889, and *Ṣaḥīḥ Muslim*, hadith #257). It is noteworthy that, even at this length, one retains the ability to pick up fine objects and make use of his nails for other tasks such as scratching, which are discussed in detail in the following footnote. (MAK)

lv An itch has many similarities to pain. Both are unpleasant stimuli, yet they evoke different behavioral response patterns. Pain results in a withdrawal reflex from the stimulus, whereas an itch leads to a scratch reflex. The two systems both use the same unmyelinated nerve fibers, but information for each is conveyed in two distinct systems that use the same peripheral nerve bundle and central spinothalamic tract. It has also been hypothesized that motivational aspects of scratching, including the frontal areas of the brain, are associated with reward and decision making, thus explaining the pleasure that is often felt when one scratches. Once again, the author astutely observes and notes this, marveling at how all of this is a mercy of God ﷻ towards creation [McMahon: pp. 219-27]. (KMR)

lvi Fingernails are primarily composed of keratin that makes them harder than skin, which is primarily composed of softer proteins such as collagen. At the same time, they are softer than bone [Bolognia: pp. 980-81]. (KMR)

lvii Here, the author takes itching one level deeper, as he has done with other anatomy in this treatise. He states that another Divine Mercy is that even if one cannot hone in on the exact source of itching, one can use his fingernails and target the general area to provide relief. Remarkably, this is all done subconsciously.

This observation is further strengthened by a Prophetic statement encouraging the believer, upon awakening, to wash his hand separately prior to making ablution, for "he does not know where his hand spent the night (*fa innahu lā yadrī ayna bātat yaduhu*)" (*Muwaṭṭa'* of Imam Mālik, Book of Purity, hadith #38).

This indicates that the hand wanders during the night when one subconsciously scratches oneself while asleep. The brilliance of the ability of al-Ghazālī ﵀ to connect a Prophetic statement with his anatomical insights must be appreciated here. (MAK)

lviii This targeting occurs under autonomic control, as one is able to do this while both awake and asleep. Human beings often scratch subconsciously while awake, allowing them to concentrate on more important tasks. While asleep, this ability to scratch allows the human being to sleep undisturbed, so that he does not have to wake up every time he has to scratch himself and suffer tiredness from interrupted sleep. God ﷻ therefore created the mechanisms to allow for uninterrupted and restful sleep: "And We made your sleep for resting" (*al-Naba'*, 9). (KMR, MAK)

lix Perhaps this explains why, when one has an itch, no other tool brings about a level of relief and comfort as that brought about by one's fingernails. (KMR)

lx The author once again juxtaposes the beauty and strength of toes, just as he did previously with fingernails. His paradigm for human anatomy at this level is to invite the reader to appreciate both form (beauty) and function (strength) in physical creation. (MAK)

lxi The author wisely notes that bones are of different types, based on their function and location in the body, and that these different forms are necessary for mechanical functions. He implies that if bones were all of the same type, complex and uniform movements would not be possible despite the bones being made of the same substances. Therefore, each bone was built upon a form related to its need – form following function is the theme here. For example, weight-bearing bones such as the pelvis and femur are thick and broad; the bones of the fingers are long and hollow, allowing for fine movements; the hyoid bone is unique in that it has no direct tendinous attachments to other bones, hanging freely yet supporting many neck muscles; and the bones of the face and nasal cavity are delicate, supporting numerous muscles and allowing for the gamut of facial expressions [Moore and Agur: pp. 315-24, 413-16, 500, 574, 593, 606]. (KMR)

lxii Bone marrow constantly produces red blood cells, white blood cells, and platelets, which function to carry oxygen to tissues, protect the body from pathogens, and produce clots and minimize bleeding, respectively. The bone marrow stroma is an environment where a symphony of hematopoiesis occurs; many different cells, such as fibroblasts, macrophages, and adipocytes come together under the influence of colony-stimulating factors to produce the necessary blood cells.

The stroma also contains marrow stromal cells, which are multipotent stem cells capable of differentiating into a variety of cell types. This cell type has been the recent subject of much excitement and controversy in the international research community [Kierszenbaum: pp. 155-63]. (KMR)

lxiii The author's commentary on how bones complement each other such that the convexity of one bone fits within the concavity of another bone, and how tendons and ligaments precisely fit into other bones and places of attachment, further suggests that he had intimate and exquisite knowledge of anatomy. He may have dissected or seen dissected bodies in order to convey such accurate details. This level of objective empiricism strongly suggests that this treatise was not written theoretically, but rather was a result of observations and deductions. (KMR)

lxiv A fascinating observation that one can move a single limb or portion of the body while the rest of the body remains immobile. (KMR)

lxv Modern anatomists agree that there are 29 bones in the head: 14 comprising the face, eight comprising the cranium, six inside the ears, and one at the top of the throat. It remains unclear how he arrives at 55 bones. We can only speculate that the definition of a "bone" may have been different in his time. (KMR)

lxvi The author subtly points towards symmetry as being part of Divine Wisdom and Magnificence in creation, and how this symmetry allows for balance. He specifically chooses the head because of its circular shape. It is difficult to balance a spherical object on a pole, but this is the precise arrangement of the skull resting atop the spine; stable enough for the human to keep his head still, yet lax enough to allow for movement in a multitude of angles and directions. (MAK)

lxvii Modern anatomy numbers these as eight bones. (KMR)

lxviii There are 32 teeth, so adding this to the 29 bones of the head equals 61, still not equaling his count of 55. It may be that the definition of a complete "bone" in his time differed from our modern understanding. Nonetheless, his understanding of different teeth (incisors, canines, molars, etc.) and their function for grinding and cutting food is impressive. (KMR)

lxix His count here corresponds to the modern count. (KMR)

lxx Cervical vertebrae are generally small and delicate, prone to fracture during traumatic impact. Even amongst these seven bones, the C3-C6 vertebrae differ from the C1, C2, and C7 vertebrae. C3-C6 vertebrae possess small bodies and are broader horizontally than antero-posteriorly. They have short spinous processes that are often split; the laminae are narrow and thinner superiorly than inferiorly; and the superior and inferior articular processes of neighboring vertebrae often fuse on either or both sides to form an articular pillar – the author has accurately described these in the text. The C1 (atlas) and C2 (axis) form the joint connecting the skull and spine. C1 is unique in that it has no body, because its body has been fused with C2, which forms the pivot upon which C1 rotates. Its body is deeper on the front than behind – such arrangement allows for the common movements of the head in the four cardinal directions. Even more unique is that, regardless of the length of their necks, all mammals (except sloths) possess seven cervical vertebrae [Moore and Daley: pp. 482-88]. (KMR)

lxxi i.e., seven cervical, twelve thoracic, and five sacral vertebrae, comprising 24 bones in total. (KMR)

lxxii There is a difference of opinion among modern anatomists as to whether the coccyx is comprised of three, four, or five fused bones. Al-Ghazālī ﷺ considers the coccyx as being composed of six bones, so his count differs. Again, his definition of "bone" may have been different from our modern definition [Moore and Agur: p. 213].

(KMR)

lxxiii It is interesting to note that there is a Prophetic statement that says the coccyx is the only bone that does not decompose in the grave (al-Bukhārī ﷺ, *Ṣaḥīḥ al-Bukhārī*, hadith #4814). It is also fascinating that the coccyx is termed dhanab in Arabic, which is etymologically similar to dhanb (sin), so as to imply that sins do not decompose by themselves unless active repentance is made. (MAK)

lxxiv There are 206 bones in the adult human body according to the modern anatomical count. The author is therefore off by 42 bones, but since he counts the 32 teeth as bones, we may say he is only off by ten. For an eleventh-century polymath, this is remarkable nonetheless [Moore and Agur: p.10]. (KMR)

lxxv Even if the count is off, the author once again states his primary focus here. (MAK)

lxxvi There are several Qur'anic verses that encourage man to look at himself and appreciate the Divine Majesty and Mercy that created him. Two verses merit mentioning here: "Soon We shall show them Our Signs in the farthest expanses of the universe and within themselves until it will be manifest unto them that this is the Truth…" (*Fuṣṣilat*, 52) and "And (there are Signs) also in yourselves; will you then not see?" (al-Dhāriyāt, 21). Of course, this pales in comparison to the splendor of the heavens and Earth: "Certainly, the creation of the heavens and earth is a far greater matter than the creation of mankind, but most of mankind knows not" (*Ghāfir*, 57). Al-Ghazālī ﷺ will fully elucidate this theme in the next several pages. (MAK)

lxxvii The modern anatomical count is either 639 or 640 muscles. (KMR)

lxxviii There are six intrinsic muscles within the orbit, and three extraorbital muscles – the corrugator supercilii, procerus, and frontalis muscles – that help each eye to move, adding up to 18 total muscles that help move the eyes and eyelids bilaterally [Moore and Agur: pp. 538-41]. A decreased functionality in any of these muscles can cause problems such as strabismus and ptosis, which may cause amblyopia (irreversible loss of vision), especially in young patients. (KMR)

lxxix It is worth mentioning here that he seems to classify the body compositionally (bones, muscles), whereas we classify the body regionally (organs, limbs) in modern times. Hence, under this discussion of muscles, he returns to using the eye muscles as an example. (KMR)

lxxx His statement here can mean several things. Given the corpus of human knowledge, instruments, and techniques available during his time, further anatomical and physiological understanding of these body parts would have been difficult. Through the development of modern technology and methods, science has been able to elucidate further detail of these tissues, down to the microscopic and molecular level. (KMR)

lxxxi As stated in the Introduction, on another level (compared to the above footnote), the author hints that there are metaphysical lessons to be learnt when one ponders the existence of these tissues and the Divine Wisdom contained therein. He implies that such knowledge cannot be gleaned from mere physical examination; it requires the disciple to use his mystical insights to gain these lessons. This idea fits well with the hierarchy of epistemology of al-Ghazālī ﷺ, with physical examination and experimentation being subservient – or at best, a foundation – for knowledge of realities. He says this is a result of specific and disciplined practices, culminating in Divinely inspired knowledge (ilhām), the pinnacle of human knowledge. (MAK)

lxxxii Due to special neural-mediated reflexes that occur between the brain and cardiovascular system over the span of seconds, arterial pressure is kept relatively constant when we transition from sitting to an upright position. Standing causes our blood pressure to drop from the effect of gravity, which can lead to decreased perfusion of the brain and cause fainting (syncope). To prevent this, special receptors called baroreceptors located in the walls of the carotid sinus and aortic arch (common carotid artery and the aorta) function as mechanical receptors sensitive to changes in pressure or stretch due to increases or decreases in arterial pressure within these vessels. This information is then relayed to the cardiovascular vasomotor centers in the brainstem that alter the outflow of the autonomic nervous system (i.e., to produce more sympathetic and less parasympathetic stimulation) to the cardiovascular system to increase heart rate and contractility. This in turn will increase cardiac output (blood flow) and increase total peripheral resistance through constriction of arterioles and veins, resulting in a compensatory increase in arterial pressure to prevent decreased perfusion when one stands [Boron: pp. 535-46]. (KMR)

lxxxiii The word *'ilāj* linguistically means treating or handling. It also has the possible meaning of "healing", which is certainly one endeavor for which hands are necessary. (MAK)

lxxxiv It is quite marvelous that throughout this head-to-toe anatomical survey of the human body, he often takes "breaks" to simply marvel at the intricacy and sophistication of the creation of the human being as a vehicle to marvel at the Creator ﷻ. The Muslim professional may similarly wish to take "breaks" in his work to similarly marvel at the Creator ﷻ. (MAK)

lxxxv He means to say that the height and length of one's limbs are fixed by adulthood and eating more food does not change this. An increase in his limbs in particular would be most unwieldy and awkward for man. (MAK)

lxxxvi The stance of al-Ghazālī ﷺ on over-eating has been well-documented, perhaps most notably in a book titled Breaking the Two Desires in his brilliant magnum opus The Revival of the Religious Sciences (*Iḥyā' 'Ulūm al-Dīn*). (MAK)

lxxxvii For example, the stomach has a limit on the amount of food it can handle and how much it needs to be satiated. Exceeding this limit can cause indigestion and a myriad of other long-term complications. The author marvels that there is a mechanism through which the human being may know he is sated and thus not cross this boundary. (MAK)

lxxxviii The progression of these rhetorical statements and challenges is al-Ghazālī ﵀ at his prototypical best, and is an indirect reference to the famous verse: "Indeed the creation of the heavens and the earth are greater than the creation of mankind, but the majority of mankind knows not" (*Ghāfir*, 57). (MAK)

lxxxix One may say that humans are capable of appreciating the intentions of God ﷻ, but are not capable of appreciating all of the Divine Wisdom behind creation. (MAK)

xc *al-Nāzi ʿāt*, 27. (MAK)

xci A reference to *al-Taghābun*, 3. (MAK)

xcii The skeletal system begins to develop during the end of the third week of gestation (approximately day 22). Two longitudinal columns of paraxial mesoderm arise laterally to the notochord and neural tube (the future vertebral column and central nervous system) and become segmented into somites that differentiate to form future vertebrate, ribs, muscle, and dermis. Limb buds appear from mesenchymal cells in the lateral mesoderm toward the end of the fourth week. These limb buds elongate during the fifth week, and condensations of mesenchymal cells aggregate to form bone models. Flat bones develop from these bone models located within preexisting membranous sheaths (via intramembranous ossification), while long bones (limb bones) develop from mesenchymal bone models transformed into cartilage bone models (which form during the fifth week). These later become ossified (via endochondral ossification) at the end of the embryonic period (around the end of the eighth week).

At birth, the diaphysis (shaft) is mostly ossified, while the epiphysis (ends) is still cartilaginous to allow further elongation of the bone until the epiphysis and diaphysis fuse at approximately 20 years of age. His use of the word "defer" is quite accurate, as by this week of embryonic development, major organs and systems, such as the lungs and central nervous system, have already begun to develop. The heart begins to beat by day 22-23; this is the first functional embryonic to form. Deferring the appearance of the skeleton allows the embryo to utilize nutrients and energy towards the development of more essential organ systems. This delay also keeps the fetus relatively "lighter" during the initial pregnancy, thereby lessening the burden on the mother to a degree [Moore and Persaud: pp. 3, 383-85, 410]. (KMR)

xciii The author uses the word *ʿaṣab*, an inclusive term for nerves, arteries, and veins. I have left the word as "sinew", which broadly refers to a fibrous tissue that joins one body part to another and can include bones, tendons, ligaments, etc. The author's intent may be that just as blood flows through vasculature, food flows through the stomach. But it is also to indicate that the stomach is a connector organ in the middle of the digestive system.

xciv The word *ʿaṣab* in pre-modern Arabic referred to a sinew or something that was flexible yet strong. This definition seems to appropriately describe the stomach, as it can shrink and elongate accordingly, yet still maintain enough mechanical and chemical strength to allow for digestion. (MAK)

xcv The stomach is capable of digesting food only after it has passed through the oral cavity. The author once again describes the physiology of this mechanism with exquisite detail. The stomach is surrounded by both parasympathetic (stimulatory) and sympathetic (inhibitory) plexuses (anterior gastric, posterior, superior and inferior, celiac, and myenteric) that regulate processes such as the motor activity of stomach muscles, secretory processes of stomach cells, and the rate of digestion itself through the opening and closing of the esophageal and pyloric sphincters. Through these complimentary activities, the stomach engages in digestion only where there is a need; it does not unnecessarily waste energy when there is nothing to digest [Guyton: pp. 720-33]. (KMR)

xcvi The process of chewing mechanically breaks down food into a bolus that may pass more easily from the oral cavity to the esophagus. It also facilitates the mixing of saliva with the food bolus to begin the initial stages of digestion. This has been discussed previously. (KMR)

xcvii The author observes that different tissues of the body require different forms of energy. This correlates with our present understanding. For example, the brain requires large amounts of glucose, whereas bones require minerals such as calcium and phosphate. (KMR)

xcviii The primary components (65-95%) of hair are keratins, which are proteins made up of long polymers of amino acids. Hair keratin contains a large amount of sulfur (5%), which is a necessary component of cysteine, a key amino acid required for the formation of keratin. No other type of cell in the human body contains such a large amount of sulfur. Hair also contains trace amounts of arsenic, chromium, iron, and magnesium [Bolognia: pp. 734, 966-67]. (KMR)

xcix Humorism, a philosophy of Greek and Roman physicians, remained the prevailing view of the human body even among Muslim physicians in the 11th century. The theory proposed that the human body was filled with four humors: – black bile, yellow bile, phlegm, and blood. Only when these were in balance would a person remain healthy. Diseases were considered to be a result of an excess or deficiency in one or more of these humors. In modern times, we have the luxury of passing off this theory as outdated, but it is also worth noting that the author ascribes to the "most current" anatomy theory of his particular time and bases his own anatomical survey of the human body around that theory. (KMR)

c The internal cavity of the bladder is lined with transitional epithelium, which allows the bladder to hold urine that contains concentrated amounts of toxic waste and metabolic byproducts that would damage other cells of the body. Transitional epithelium is a specialized epithelium found only in the bladder, ureters, and urethra. It is unique as it can tolerate both mechanical and chemical trauma, as well as stretch to accommodate and facilitate both the storage and passage of urine. As the bladder fills with urine, the transitional epithelium transforms from dome-shaped cells to a more flattened epithelium, thereby allowing the volume of the bladder to increase in response to even minimal increases in intravesical pressure [Kierszenbaum: pp. 419-20]. (KMR)

ci This is to say that the fetus has no need to call out for assistance, nourishment, or protection while in the womb. This is juxtaposed with the newborn, who calls out for assistance, but does not specify what is needed at that particular time. The author is encouraging the reader to think about how his needs were met in these two stages – one that did not require any "calling", and one that incorporated an "unspecified" calling out. (MAK)

cii For example, even in the womb, the final stages of lung development do not occur until just before delivery, as the fetus has no need for them while in the womb. Another example is the patency of the ductus arteriosus, a vessel that connects the pulmonary artery to the aorta. This allows oxygenated blood received from the mother to bypass the immature lungs and feed directly into the aorta and the rest of the body. The ductus arteriosus closes shortly after the baby is delivered. (KMR)

ciii Tooth development occurs as early as the sixth week, with the formation of tooth buds (dental laminae) out of a thickening of oral epithelium. However, these buds do not erupt with teeth until six to eight months of age with the appearance of medial incisors, continuing until 20-24 months when the second molars appear [Moore and Persaud: pp. 494-95]. (KMR)

civ Since the newborn is unable to digest solid food for the first six months, teeth at this stage of life would not be of any benefit. As the digestive system matures by six months after birth, teeth develop concurrently so that the child can consume solid food. (KMR)

cv This is an extremely subtle yet astute observation. The newborn has to deal with multiple biological issues necessary for survival outside the womb, such as learning how to breathe and feed. The fetus is not aware of his surroundings; certainly, the developed mechanisms of cognition are not in place. But being aware would actually yield him no benefit. In fact, it may be argued that had the fetus been aware, such a state may be harmful as a result of multiple stimuli or a sense of claustrophobia. If the newborn immediately possessed the ability to ponder and contemplate, he would become so absorbed in trying to make sense of this new world that the tasks of breathing and feeding would be ignored. Thus, intelligence and awareness develop gradually in a stepwise manner that is predictable in most infants. (KMR)

cvi This refers ostensibly to both ease in creation and ease in inventiveness being manifestations of Divine Power and Creativity, respectively. (MAK)

cvii The unique physiology of fetal circulation enables the fetus to survive in the womb. During prenatal life, the fetal cardiovascular system is structurally and operationally different from postnatal life. This is because fetal lungs have no respiratory function, since air is unavailable to them. As mentioned previously, the lungs remain nonfunctional while the function of gas exchange is handled by the placenta where fetal and maternal vessels exchange substances and nutrients. Oxygenated blood flows from the placenta into the right side of the heart. A small amount of blood from the right atrium follows the usual pathway through the pulmonary artery to the lungs, while a large percentage of blood passes from the right atrium directly into the left atrium through a passage called the foramen ovale. This oxygenated blood is then pumped out of the left ventricle through the ascending aorta to perfuse the upper half of the body. The oxygenated blood passing through the pulmonary artery en route to the lungs is mostly diverted to the descending aorta via the ductus arteriosus to perfuse the lower half of the body. Although this system is less effective than lungs that are directly exposed to the atmosphere, fetal blood consists of fetal hemoglobin, which has a higher affinity for oxygen compared to adult blood. This allows for better extraction of oxygen from maternal blood and aids in transport of oxygen across the placenta to the fetus. Following birth and separation from the umbilical cord, there is increased pressure in the aorta and subsequent vasculature that allows the lungs to expand, the foramen ovale to close, and the ductus arteriosus to constrict and close up. These events help establish the postnatal circulatory system necessary for survival outside the womb [Rhoades: pp. 647-54]. (KMR)

cviii Post-term pregnancy is defined as a pregnancy that extends beyond 42 weeks. Recent advances in fetal monitoring and imaging have led to a greater awareness of the risks of such pregnancies. Fetal and neonatal mortality rates sharply increase after 40 weeks, and antepartum stillbirths account for more perinatal deaths than either complications of prematurity or sudden infant death syndrome. The physical and emotional risks to the mother are also quite considerable yet underappreciated. Maternal morbidity also increases after 40 weeks of gestation [Cotzias: pp. 287-88; ACOG Practice Bulletin: pp. 639-46; Caughey: pp. 57-62]. (KMR)

cix Maternal production of breast milk and how it suffices the newborn is one of the most fascinating occurrences in nature. Human breast milk contains protein, lactose, fat, and minerals, such as calcium and phosphate. These components not only provide adequate nutrition for the newborn but are also essential building blocks for the growth and development of the infant. Breast milk also helps protect the infant from infections.

For example, there are large quantities of IgA that confer passive immunity against mucosal infections (from gastrointestinal and respiratory pathogens) until the infant can produce his own immunoglobulins in sufficient amounts (at approximately 3-4 months of age). Other constituents of breast milk include peptide hormones and growth factors [Kierszenbaum: pp. 655-56; Coico: pp. 48-49]. (KMR)

cx This is a reference to the verses: "Has there (ever) come upon man a period of time in which he was a thing unremembered? Surely We created man from a drop of thickened fluid in order to try him, so We make him hearing and seeing" (*al-Insān*, 1-2). (MAK)

cxi While I have inserted "the individual nutrients", this may also mean that the liver distinguishes between food and contaminants, as one of its primary functions is to filter and remove harmful toxins from the blood. (KMR)

cxii It is noteworthy that the author makes a distinction between waste and excess. One may consume moderately, but will still excrete a modest amount of waste, whereas overconsumption creates additional excreted waste. He may be suggesting that waste refers to urine, whereas excess refers to stool (i.e., the unusable portion of consumed food). Or perhaps he is drawing inspiration from a Prophetic description of Paradise, whose dwellers will not excrete any type of waste, as their bodies will completely absorb and benefit from the nutrients in consumed food. He is subtly pointing out that earthly bodies cannot benefit completely from consumed food and will excrete some amount of waste products. He marvels at how these mechanisms, which are part of Divine Mercy to creation, remove both waste and excess, i.e., that there is not one organ for removing waste and another for removing excess. (MAK)

cxiii After food is digested in the stomach, it continues to the intestines. The bulk of digestion occurs in the small intestine, particularly the duodenum. The small veins of the intestines feed into the superior mesenteric and inferior mesenteric veins (later joined by the splenic vein), which subsequently feed into the hepatic portal vein. All digested nutrients are sent to the liver via the portal vein. The significance of this vascular organization is that many of the absorbed digestion products are subject to metabolic processing and detoxification by the liver before being distributed to the cells of the body [Boron: pp. 975-76]. (KMR)

cxiv This is yet another subtle point showing his understanding of physiology. He notes that food that passes through the portal vein has been broken down for the liver to use, package, and distribute to the rest of the body. He highlights that food is broken down by the time it reaches the portal vein, but that the portal vein itself serves as a barrier to ensure that incompletely digested and potentially harmful products are not sent to the liver. The "thin and fine" description likely refers to filtering capability, not a physical description of the blood vessel [Levy: pp: 493-95, 637-38]. (KMR)

cxv These elemental components are mixed into the blood or sent via carrier molecules

– such as fats in chylomicrons – to the rest of the body. In the small intestine, products of lipid digestion are re-esterified within intestinal cells to form the original ingested fats and packaged into lipid-carrying particles called chylomicrons. These structures are composed of lipids and lined with surface apoproteins, which are essential for chylomicron absorption and subsequent dietary lipid absorption. These chylomicrons are then released into the lymphatic system because they are too large to enter vascular capillaries but are eventually delivered to the bloodstream after the thoracic duct empties into the left subclavian vein. Chylomicrons enable fats and cholesterol (which have poor solubility) to move within the blood and deliver their contents to the liver, adipose tissue, cardiac and skeletal muscle tissue for further metabolism or storage [Boron: pp. 964-67]. (KMR)

cxvi After blood is purified and passed through the liver, it exits through the hepatic veins that feed directly into the inferior vena cava, which then feeds into the right atrium of the heart. From here, this blood mixes with blood from other parts of the body and passes through the right ventricle and lungs, where it becomes oxygenated – now containing both oxygen and nutrients – then travels to the rest of the body. (KMR)

cxvii Ghāfir, 64. (MAK)

cxviii The word used in the text is *ma ʿābiḍ*, which is not found in the dictionary. This may actually be a misprint during the copying of another word: *maghāyiḍ*, which would indicate a gathering container. A simple error in transcribing the dots of these two words may explain the former word found in the versions of the text used for this translation. Under the assumption that this is *maghāyiḍ*, I have translated this as "receptacles", which would keep the intended meaning of what is discussed. This may also be similar to *aw ʿiyah* (vessels), especially since he uses this terminology in the next sentence. God knows best.

cxix Here again, the author ﵀ makes an excellent point in discussing the need for an additional component or a medium for a given sense to gain any information about the object. For example, eyesight requires light as a medium that conveys information from the object to the human being, whereas olfaction requires air to carry fragrances to the nose. He is saying that even the existence of these mediums is from Divine Wisdom, for man's senses would be incomplete and useless without them. (MAK)

cxx It may be argued that people with a deficiency in one sense are given enhanced abilities in their other senses. For example, it is now known that a person who is blind has a heightened sense of hearing as a result of visual cortex areas of the brain being used towards the faculty of listening [Stevens: pp. 10734-41]. (KMR)

cxxi It is fascinating that Muslims historically did not view blindness as a handicap. In fact, Muslim rulers considered it to be a state responsibility to assist handicapped citizens. For example, the Caliph ʿUmar ibn ʿAbd al-ʿAzīz ﵀ (d. 102/720) routinely queried the provincial rulers for a list of all handicapped citizens so the state could provide them with an employee to look after their daily affairs and assist them in attending congregational prayers. The existence of numerous blind scholars and memorizers of the Qur'an (*ḥuffāẓ*), such as the famous scholar Abū Isḥāq al-Shāṭibī ﷺ (d. 790/1388), bears witness to this. It is also one of the miracles of the Qur'an that a blind person can simply hear its verses and memorize its entire text with equal (or even better) proficiency than a person with sight. Finally, the Prophet ﷺ is reported to have said, "God, the Glorious and Exalted said: 'When I afflict My slave in his two dear things (i.e., his eyes) and he endures patiently, I shall compensate him for them with Paradise'" (*Ṣaḥīḥ Bukhārī*, hadith #5635). (MAK)

cxxii It is noteworthy that while this treatise marvels at the perfection in the creation of God ﷻ, the author concluded this paragraph by acknowledging that God ﷻ also creates "imperfection" (such as abnormalities in the body). It is crucial to remember that while benefit and balance are the norm, occasional imbalance itself is a mercy, as it is a reminder that His Bounty is not to be taken for granted. Like a bitter medicine, there is benefit for those who can understand and appreciate this. (MAK)

cxxiii This is a subtle, almost tongue-in-cheek justification for Divine Oneness by using the head as a metaphor. (MAK)

cxxiv One of the author's motifs is to comment extensively in some places while allowing the reader to reflect in others, especially on topics where the discussion is quite evident. He makes use of this brilliantly throughout this treatise. (MAK)

cxxv Here again, he uses the door analogy as previously mentioned. (MAK)

cxxvi His reference to the appearance of the gross anatomy of the brain, with its gyri and sulci, is extremely accurate [Moore and Agur: pp. 516-28]. (KMR)

cxxvii The discussion now begins to move towards incorporating spiritual and loftier aims rather than simple descriptions. As mentioned in the Introduction, the author embraces the use of scientific observation as a tool towards appreciating the Divine to a greater degree than one who does not undergo this exercise. Here, he points out that one proof for the existence of God ﷻ is contained in the barriers that protect man's brain such that he can even think in the first place. He subtly argues that if these barriers did not exist, then the brain would be harmed, and man would not be able to think. Thus, it behooves man to think about the Creator ﷻ of not only himself and his organs – including his brain – but even the barriers that protect his brain, which allow him to think and learn in the first place. As the brain allows man to perceive, and God ﷻ created it, He is therefore the source of all knowledge, both sacred and mundane: "He taught man what he knew not" (*al- 'Alaq*, 5). (MAK)

cxxviii This membrane, or the pericardium, consists of two layers: a visceral layer, adherent to the outermost layer of the heart (the epicardium), and a parietal layer. Between the visceral and parietal layers is a thin layer of fluid that provides lubrication for the continuous and smooth movement of the enclosed heart. The pericardium resists distention, thereby preventing large and rapid increases in cardiac size and sudden overdistention of the heart chambers [Levy: p. 253]. (KMR)

cxxix This is perhaps the author's most exceptional observation in the entire treatise. He is essentially describing the system of pulmonary circulation in the body, wherein deoxygenated blood from the right heart enters the lungs, becomes oxygenated, and returns to the left heart for its subsequent pumping to the rest of the body. William Harvey (d. 1657) is generally credited as the first Western physician to correctly elucidate the physiology of human circulation. However, the famous Muslim polymath Ibn al-Nafīs (d. 687/1288) had already correctly described circulatory physiology, especially details of pulmonary and coronary circulation. During the time of al-Ghazālī ﵀, the prevailing theories of Galen (d. 199 or 217) and Ibn Sīnā (d. 428/1037) proposed that blood from the right side of the heart flowed into the left side through invisible pores, where it subsequently mixed with air and became "spirit" that flowed into the rest of the body. It is noteworthy that this description of circulation by al-Ghazālī ﵀ predates both Harvey and Ibn al-Nafīs, but also corrects the faulty physiology of Galen and Ibn Sīnā. To date, no serious consideration has been given to al-Ghazālī ﵀ being the first person to develop a more scientifically accurate theory of circulation. (KMR)

cxxx The text here may be read in two ways, the other being: "Then He filled his stomach with air for his wholesomeness and other reasons." (MAK)

cxxxi This is perhaps a reference to the Prophetic statement: "Man fills no vessel worse than his stomach. It is sufficient for the son of Ādam to have a few mouthfuls to give him the strength he needs. If he has to fill his stomach, then let him leave one-third for food, one-third for drink, and one-third for air" (*Sunan Abī Dawūd*, hadith #393, *Sunan Ibn Mājah*, hadith #3349). There are numerous health benefits in keeping the stomach filled with air. (MAK)

cxxxii In Muslim architecture, the bathroom was hidden in the back part of the house so that it was not readily apparent, especially to a guest who entered through the front door. Similarly, the author indicates that the anus (and excretion of feces) is hidden between the flesh and muscles of the buttocks (gluteus muscles). Thus, the design of this excretory organ is a further honoring for the human being since it is covered and not readily observable, even to himself. (MAK)

cxxxiii Perhaps a reference to the hadith of fiṭrah: "Ten actions are part of natural behavior (*fiṭrah*): trimming the mustache, growing the beard long, using the toothbrush, sniffing water into the nose, clipping the nails, washing the knuckles, removing hair from the underarms, shaving pubic hair and cleaning the private parts with water (after the call of nature)" (*Ṣaḥīḥ Muslim*, hadith #261). (MAK)

cxxxiv The author comments on how the drive to satiate hunger and thirst is powerful and remains distinct from all of man's other faculties. Indeed, it is the strength of this drive that allows him to stay alive. Even if man is busy in other affairs and is using his awareness towards those endeavors, this distinct force reminds him and impels him to eat and drink in order to preserve his life. Anatomically, the limbic system of the brain – specifically areas in the midbrain and certain nuclei in the thalamus – is responsible for the sense of hunger and thirst, as well as their satiation. It is quite remarkable that the author notices this existence, even if he does not comment on its anatomical location. One may argue that he recognizes that different areas of man's awareness govern his specific needs and functions. (KMR)

cxxxv He notes that there are both sociological obstacles that prevent man from intimate relations and also biological ones, such as difficulties with pregnancy, childbirth, and rearing. However, the mutual desire of the two genders is stronger than any of these social, psychological, financial, and physiological obstacles; were it not for this resilient desire, human beings would not sire new generations. (MAK)

cxxxvi Such as the kidneys, liver, and intestines. (KMR)

cxxxvii Here again, the author makes a subtle point of stating that man's ability and tendency to forget is a Divine Mercy. Linguistically, the Arabic word for man (insān) is said to be partially derived from the trilateral verb *na-si-ya*, which means to forget, so as to say that it is in the nature of man to forget. The other etymological origin is said to be from uns (intimacy), so as to say that human beings require companionship and human contact. Whereas we may view forgetfulness as a universal shortcoming amongst all human beings, al-Ghazālī ﵀ suggests that forgetfulness itself is a blessing in many ways. (MAK)

cxxxviii He is referring to the period of time that passes after a given event that allows one to forget distressful thoughts from that event. (MAK)

cxxxix Perhaps a reference to one possible interpretation of the famous hadith: "Indeed among the words mankind received from Prophets are this: 'If you feel no shame, do as you wish.'" In other words, modesty is a driving factor to prevent evil from occurring, especially communally and socially. (MAK)

cxl Referring to the brain as the organ that houses this emotion of modesty.

cxli Social modesty is a uniquely Adamic feature; the physical brain allows for this powerful emotion. (MAK)

cxlii This is a reference to the Qur'anic verse, "He taught him the mode of expression" (*al-Raḥmān*, 4), as a sign of this being one of the greatest blessings of God ﷻ to mankind. It is notable that this fact is mentioned at the beginning of one of the most majestic and beautiful chapters of the Qur'an. The main theme contained therein is the repeated mention of the blessings of God ﷻ upon man and jinn, and an exhortation to remember, give thanks, and not deny these favours. The favour of speech is mentioned immediately after God ﷻ mentions the creation of man, almost as if to highlight that this blessing is the pinnacle of favours that God ﷻ gave to mankind. The author astutely inserts this Qur'anic exegesis here. (MAK)

cxliii Here he includes the ability and parts of the body that allow man to write. This is a reference to the Qur'anic verse, "He who taught (to write) with the pen" (*al- ʿAlaq*, 4). It is notable that this verse appears in the first five verses that were revealed to the Prophet ﷺ, further indicating its importance. (MAK)

cxliv The order here is interesting: speech is mentioned first, then writing skills. Muslims have historically developed both traditions, though the oral tradition is particularly esteemed, especially in regard to rote preservation of revelation and Prophetic traditions. (MAK)

cxlv The text here reads as "Romans", but "Greeks" seems more appropriate in this context. (MAK)

cxlvi He notes that man's health is dependent upon not only the physical body being in balance and proportion, but his emotional and psychological states as well. This is mentioned in certain Qur'anic verses, such as the verse previously mentioned: "Indeed, We have created all things in proportion and by measure" (*al-Qamar*, 49). (MAK)

cxlvii While scholars of Sacred Law are in disagreement over the root sin of Satan that caused him to be expelled from heaven, a strong opinion is that it was his jealousy towards Ādam, in lieu of the latter being chosen by God ﷻ. This created anger and led to arrogance, which is mentioned in several places in the Qur'an through the use of the verb istakbara, which translates to "he sought or attempted to make himself grand". The subtle point here is that arrogance certainly was Satan's crime, but the arrogance itself was comprised of anger and jealousy.

Al-Ghazālī ؒ notes that both lacking and possessing excess anger and jealousy is detrimental to the well-being of man, and one's well-being can only be safeguarded when one possesses moderation between the two and can properly display and quell these emotions at appropriate times. (MAK)

cxlviii The author is referring to ghibṭah, which is different from ḥasad. While both may be translated as jealousy, the former implies a desire for something that another person has in terms of religious bounties – knowledge, spending wealth on charitable causes, etc. The key component is that the person desiring does not wish any decrease in these blessings or misfortune upon the possessor of such bounties, nor does he believe that such a person is undeserving. For this would be tantamount to questioning God

ﷻ, who through His Divine Wisdom gave each person their respective talents and resources. Ḥasad is blameworthy envy and is a sin of the Devil. It is when a person desires from another, but believes the other person is undeserving and wishes lessening or ruin upon him. (MAK)

cxlix The theme of embracing both Divine Giving and Withholding as goodness is a universal concept in Muslim literature. It builds upon verses such as, "And God ﷻ knows while you know not" (*al-Baqarah*, 216), which establish the absolute Knowledge of God ﷻ and remind man to accept His Decree. (MAK)

cl This is to say that every generation benefits from previous generations' civilizational artifacts, relics, and ideas; each generation does not have to start anew in terms of development. (MAK)

cli This is likely a reference to the verse: "And in the earth there are tracts side by side and gardens of grapes and corn and palm trees having one root and (others) having distinct roots – they are watered with one water, and We make some of them excel others in fruit; most surely there are Signs in this for a people who understand" (*al-Ra 'd*, 4). (MAK)

clii The word here is marākib, which includes all forms of transportation, including boats. It may mean only boats and watercraft. (MAK)

cliii This is likely also a reference to the verse: "[He made] horses and mules and asses that you might ride upon them and as an ornament; and He creates what you do not know" (*al-Naḥl*, 8). (MAK)

cliv His mentioning of birds here is a subtle hint to the next treatise that deals with the Divine Wisdom in the creation of birds in a similar fashion. Inspired by a Qur'anic verse, he proceeds to observe a variety of birds to arrive at specific conclusions. His scientific method is precisely elucidated in this treatise, and is recommended reading for those who wish to better understand his epistemology of science. (MAK)

clv This likely refers both to the different types of clothing found in different communities, as well as the variety of clothing that can be made from different raw materials; these have also been created and given to man to use freely as Divine Favors. This is reinforced in verses such as, "Do you not see how God ﷻ has made subject to you all that is on the earth…" (*al-Ḥajj*, 65), and, "Do you not see that God ﷻ has subjected to your (use) whatsoever is in the heavens and whatsoever is in the earth, and has made His Bounties flow to you in exceeding measures (both) outwardly and inwardly?" (*Luqmān*, 20). (MAK)

clvi The author mentions "in him" in the wake of mentioning many outward blessings because he subtly notes that none of these outward blessings could be enjoyed by man if he did not possess the inner faculties, organs, and awareness to do so. (MAK) clvii This is likely a reference to the verse: "God ﷻ has favored some of you above others in provision; those who are more favored are by no means going to hand over their gifts to those whom their right hands possess so that they (too) should be equal in that respect. Is it, then, the Grace of God ﷻ that you deny?" (*al-Naḥl*, 71). (MAK)

clviii i.e., whether they are poor or rich. (MAK)

clix He says that those who blindly follow their whims are slaves to their lusts, which echoes a Qur'anic passage in that same theme: "Have you not seen the one who takes his own whims (and lusts) as his god? Will you then be a guardian over him? Or do you think that most of them hear or understand? They are only like cattle – nay, but they are even worse astray!" (*al-Furqān*, 43-44). (MAK)

clx al-Isrā', 70. (MAK)

clxi The author ﷺ suggests that any wisdom and trustworthiness found in human beings are themselves a minute fraction from the Divine Wisdom and Perfection of God ﷻ; these have been placed in man by Him d. (MAK)

clxii al-Dhāriyāt, 21. (MAK)

clxiii He marvels that man's intelligence and ability to be aware of himself is enough proof of his own existence. This is a profound statement of the French philosopher, René Descartes (d. 1650), famous for and generally credited with developing an ontology to prove the existence of man: Cogito, ergo sum ("I think, therefore I am"). Descartes argued that the first principle is that thought exists, and since thought cannot be separated from myself, therefore I too must exist. Descartes maintained the corollary to be true as well: if I am doubting, then someone must be doubting, so that too proves my existence. Essentially, one may understand that whenever one wonders if he exists, that act itself is enough proof that he exists. The statements of al-Ghazālī ﷺ in the text are the same. However, he predates Descartes in arriving at this ontological proof by several centuries. Substantial scholarly works have discussed Descartes' philosophical and academic "borrowing" of al-Ghazālī ﷺ, and it is certainly plausible that Descartes had access to the works of al-Ghazālī ﷺ and took liberally from them (Najm: 133-41; see also Mahmoud Bejou's introductory notes on al-Ghazālī's Munqidh, published in Damascus, 1992). (KMR)

clxiv The author juxtaposes the above discussion regarding the power of man's intelligence by mentioning that despite the power of the intellect in proving man's existence and being itself a proof for this, it is utterly incapable of describing or understanding its Creator d – the reader should reflect upon this powerful paradox. (MAK)

clxv This suggests that God ﷻ has given man an intellect followed by the ability and power to learn, as found in, "(He is) the One who taught with the pen. He taught man what he knew not" (*al- ʿAlaq*, 4-5). It may be further understood that man can learn for the duration of his life, for the Prophet ﷺ instructed his followers to seek knowledge "from the cradle to the grave". Al-Ghazālī ؒ notes that the internal mechanisms to receive the instruction of God ﷻ and to subsequently follow the Prophetic command are also present in the human being until he dies. (MAK)

clxvi This is remarkable, as he is comparing the physical abilities of the five senses having an inability to perceive the intellect; yet the intellect dominates these five senses both in processing sensory information and directing use of these five senses towards a given purpose. (MAK)

clxvii Man does not seek to have personification of his own mind, yet he asks for personification of God ﷻ. Those with knowledge realize: "Why do I ask for the personification of God ﷻ if I do not ask for the personification of my mind? I believe in it while it has no form or shape, but yet I do not believe in God ﷻ?" (MAK)

clxviii He concludes here that despite the lack of objective or sensory proofs, the intellect can have faith and see with this "light" until man becomes more certain about God ﷻ and the unseen than any objective thing that he sees using his physical eye. He argues that the intellect's vision is stronger than physical vision. This is a powerful idea, as he spent a considerable time earlier in the treatise marveling at the sophisticated creation and ability of the physical eye. (MAK)

clxix This is somewhat of a paradox, as he rhetorically asks which comes first: the desire towards obedience to God ﷻ or the desire towards haste in obedience to God ﷻ. It is unclear which he favors. (MAK)

clxx The author posits that despite man's potential to recognize God ﷻ, he is still incapable of describing and knowing details about his own self. Thus, man can only acknowledge his own existence elementally, as mentioned previously in the discussion on Descartes. As a further distinction between man and God ﷻ, the author adds here that God ﷻ is capable of describing Himself and possesses Divine Knowledge about Himself. (MAK)

clxxi This is an interesting point, especially given the above discussion that both the ability to remember and the ability to forget are Divine gifts that are needed at specific times. He notes that these same faculties are not under complete control of the human being – man cannot always willingly remember or forget; this subtle incapability further shows man how dependent he is upon his Creator d. (MAK)

clxxii This would literally translate as, "he would arrive upon the crossing of safety (warada mawrid al-salāmah)". I have taken some additional liberties with this phrase in order to convey its complex meaning.

clxxiii This is a subtle reference to the verse: "Have they not traveled in the land so that they should have hearts with which to understand, or ears with which to hear? Surely, it is not the eyes that are blind, but blind are the hearts of the breasts" (*al-Ḥajj*, 46). (MAK)

clxxiv The author ﷺ notes here that the intellect (*'aql*) is so powerful that not only can it be used for issues such as recognizing one's own existence and marveling at the Signs of God ﷻ, it can also be employed for more worldly matters. He marvels at the power of this faculty, for it encompasses the most holy and the most mundane of man's affairs. (MAK)

clxxv His observations here may be understood as akin to the modern fields of history, sociology, and anthropology. Even in these areas of study, he must use his intellect to learn what is beneficial and wholesome from other people and those before him, and to leave aside what is detrimental to his own well-being in his respective time and space. (MAK)

clxxvi As stated in the Introduction, his epistemology regards Prophetic knowledge as superior to intellect. However, he recognizes the usefulness of the latter. This entire treatise resulted from the combination of the two and is a proof for the necessity to embrace and amalgamate these two types of knowledge. (MAK)

clxxvii The Sun appears to give different intensities of light depending on its position in the sky and the season. (KMR)

clxxviii The concept of ru'yah. According to the author, dreaming is a faculty exclusive to human beings; it may occur during sleep or even during half-awake/half-asleep states. (MAK)

clxxix This is an amazingly profound point that he waits until the end to bring about, perhaps showing that man's exclusive ability to dream is the highest favour amongst all of the favours that have been enumerated in this treatise. Dreams can be warnings about the future or the past, so that when man awakens, he can appropriately save himself to thwart impending misfortune or rectify himself to atone for past transgressions. On another level, one may also argue that dreams are the highest of the favours of God ﷻ, as they point out the limitations of the knowledge that man can access only through his intellect. Perhaps more significantly, pleasant dreams comfort and solace human beings and can allay fear and anxiety. Finally, it is also noteworthy that the Prophet ﷺ said that true dreams are one of 46 parts of Prophethood. He ﷺ also experienced true dreams for six months before the first revelation, and his Prophethood lasted for 22.5 years, so the hadith is also mathematically correct. This is the highest level of his epistemology, as he once again emphasizes the intimate relationship between action and knowledge – knowledge should stimulate man towards actions that are pleasing to God ﷻ. Only then has he put knowledge towards its Divinely mandated purpose and made himself eligible to receive even more knowledge. (MAK)

Bibliography

1 "ACOG Practice Bulletin: Clinical Management Guidelines for Obstetrician-Gynecologists, Number 55, September 2004 (Replaces Practice Pattern Number 6, October 1997." *Obstetrics and Gynecology* 104.3 (2004): 639-46. Web. 21 Dec 2009.

2 Anderson, James. *Grant's Atlas of Anatomy.* 7th ed. Hamilton, Ontario: The Williams & Wilkins Company, 1978. Print.

3 Bolognia, Jean, Jorizzo, Joseph and Ronald Rapini. *Dermatology.* 2nd ed. Spain: Mosby Elsevier, 2008. Print.

4 Boron, Walter, and Emile Boulpaep. *Medical Physiology.* China: Saunders, 2003. Print.

5 Caughey, Aaron, and Thomas Musci. "Complications of Term Pregnancies Beyond 37 Weeks of Gestation." *Obstetrics and Gynecology* 103.1 (2004): 57-62. Web. 21 Dec 2009.

6 Chai, Tuu-Jyi, and Toby Chai. "Bactericidal Activity of Cerumen." *Antimicrobial Agents and Chemotherapy* 18.4 (1980): 638-41. Web. 21 Dec 2009.

7 Coico, Richard, Sunshine, Geoffrey, and Eli Benjamini. *Immunology: A Short Course.* 5th ed. USA: John Wiley & Sons Inc., 2003. Print.

8 Cotzias, Christina, Paterson-Brown, Sara, and Nicholas Fisk. "Prospective Risk of Unexplained Stillbirth in Singleton Pregnancies at Term: Population Based Analysis." *BMJ* 319.7 (1999): 287-88. Web. 21 Dec 2009.

9 Guyton, Arthur and John Hall. *Textbook of Medical Physiology.* 10th ed. Chile: Saunders, 2000. Print.

10 Kierszenbaum, Abraham. *Histology and Cell Biology: An Introduction to Pathology.* 2nd ed. Canada: Mosby Elsevier, 2007. Print.

11 Jahn, Anthony and Joseph Santos-Sacchi. *Physiology of the Ear.* 2nd ed. Canada: Singular, 2001. Print.

12 Kumar, Vinay, Abbas Abul, and Nelson Fausto. *Robbins and Cotran Pathologic Basis of Disease.* 7th ed. China: Elsevier Saunders, 2005. Print.

13 Levy, Matthew, Koeppen, Bruce, and Bruce Stanton. *Berne & Levy: Principles of Physiology.* 4th ed. China: Mosby Elsevier, 2006. Print.

14 McMahon, Stephen and Martin Koltzenburg. *Wall and Melzack's Textbook of Pain.* 5th ed. China: Elsevier, 2006. Print.

15 Moore, Keith and Anne Agur. *Essential Clinical Anatomy.* 2nd ed. Baltimore: Lippincott Williams & Wilkins, 2002. Print.

16 Moore, Keith and Arthur Dalley. *Clinically Oriented Anatomy.* 5th ed. Baltimore: Lippincott Williams & Wilkins, 2006. Print.

17 Moore, Keith, and T.V.N. Persaud. *The Developing Human: Clinically Oriented Embryology.* 7th ed. China: Saunders, 2003. Print.

18 Najm, Sami. "The Place and Function of Doubt in the Philosophies of Descartes and al-Ghazālī." *Philosophy East and West* 16.3-4 (1966): 133-41. Web. 21 Dec 2009.

19 Perry, Eldon and Anna Nichols. "Studies on the Growth of Bacteria in the Human Ear Canal." *The Journal of Investigative Dermatology* 27.3 (1956): 165-70. Web. 21 Dec. 2009.

20 Rhoades, Rodney, and Richard Pflanzer. *Human Physiology.* 3rd ed. USA: Saunders College Publishing, 1996. Print.

21 Riordan-Eva, Paul and John Whitcher. *Vaughan and Asbury's General Ophthalmology.* 17th ed. USA: McGraw-Hill Medical, 2007. Print.

22 Sahler, Olle, and John Carr. *The Behavioral Sciences and Health Care.* 2nd ed. USA: Hogrefe & Huber Publishers, 2007. Print.

23 Stevens, Alexander, et al. "Preparatory Activity in Occipital Cortex in Early Blind Humans Predicts Auditory Perceptual Performance." *The Journal of Neuroscience* 27.40 (2004): 10734-41. Web. 21 Dec 2009.

Table 1

COMPARISON OF BONE AND TEETH

	TEETH	BONE
Embryonic Origin	Ectoderm, mesoderm, and neural crest cells	Mesoderm
Components	Enamel, dentin, pulp, and cementum	Osseous tissue, marrow, endosteum, and periosteum
Notes	Enamel is the hardest and most highly mineralized substance found in the body, with a 95% composition of hydroxyapatite and a 5% composition of amelogenin and enamelin proteins. The dentin that lies just under the enamel is considered the second strongest substance found in the human body.	Osseous tissue is relatively hard due to a 70% mineral composition of hydroxyapatite and other minerals, but unlike teeth, it is also elastic due to type I collagen protein.

Figure 1

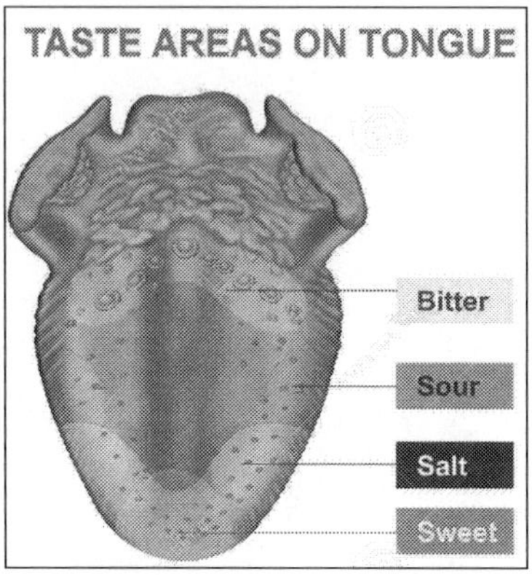

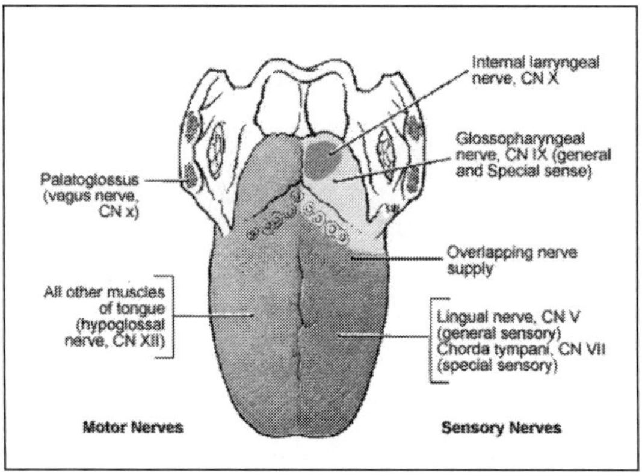

Anatomical distribution of sensory and taste regions of the tongue. In the anterior two-thirds of the tongue, sensory information is conveyed by the lingual branch (V_3) of the trigeminal nerve (CN V), whereas taste information is conveyed by the chorda tympani branch of the facial nerve (CN VII). In the posterior two-thirds of the tongue, both sensory and taste information is conveyed by the glossopharyngeal nerve (CN X).

Figure 2

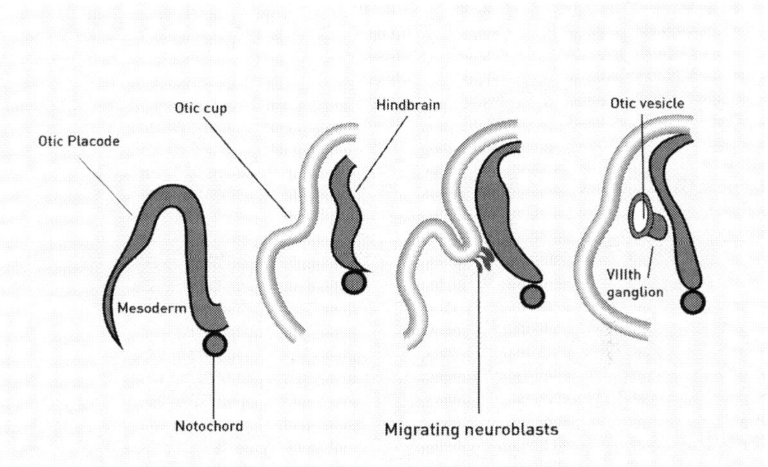

Schematic diagram representing embryologic stages of otic vesicle development. Note the process of invagination depicted in the diagram as compared to the description in the text.

SHEIKH MOHAMMED AMIN KHOLWADIA

Sheikh Mohammed Amin Kholwadia – popularly known as *Sheikh Amin* – is an acclaimed Muslim scholar and theologian based in the Greater Chicago area. The Sheikh is trained in various Islamic sciences such as Qur'anic exegesis, hadith transmission, Islamic law, and theology. Since his arrival in Chicago in 1984, he has served as a Muslim scholar in various capacities – such as being an advisor for Muslim schools, Muslim organizations, and the Council of Religious Leaders of Greater Chicagoland. Sheikh Amin is the co-author of a book on Islamic Finance entitled *Islamic Finance: What It Is and What It Could Be*. He has also written a book on Qur'anic exegesis entitled *A Spark From the Dynamo of Prophethood*. Sheikh Amin founded Darul Qasim in 1998, an institute of higher Islamic learning that facilitates intermediate and advanced levels of religious study.

DR. KAMRAN RIAZ

Dr. Kamran Riaz has a uniquely indigenous background in Islamic studies. Before high school, he completed the memorization of the Qur'an within nine months. He attended the University of Illinois in Chicago (UIC) for undergraduate studies and graduated *magna cum laude* from the prestigious GPPA-Medicine program with dual degrees in Economics and Middle Eastern History. He obtained a doctorate of medicine degree from the UIC College of Medicine and completed his ophthalmology training at Northwestern University and the University of Texas Southwestern Medical Center (Dallas). He has been a student at Darul Qasim under Sheikh Amin since 2000. He currently resides in Oklahoma City where he is an Associate Professor of Ophthalmology at the Dean McGee Eye Institute (University of Oklahoma). He and his wife are blessed with three children.

Made in the USA
Monee, IL
15 September 2025

af4f2574-cd09-48a0-bc42-4d2d82565151R01